IMMUNOLOGICAL TOLERANCE

The Novartis Foundation is an international scientific and educational charity (UK Registered Charity No. 313574). Known until September 1997 as the Ciba Foundation, it was established in 1947 by the CIBA company of Basle, which merged with Sandoz in 1996, to form Novartis. The Foundation operates independently in London under English trust law. It was formally opened on 22 June 1949.

The Foundation promotes the study and general knowledge of science and in particular encourages international co-operation in scientific research. To this end, it organizes internationally acclaimed meetings (typically eight symposia and allied open meetings, 15–20 discussion meetings, a public lecture and a public debate each year) and publishes eight books per year featuring the presented papers and discussions from the symposia. Although primarily an operational rather than a grant-making foundation, it awards bursaries to young scientists to attend the symposia and afterwards work for up to three months with one of the other participants.

The Foundation's headquarters at 41 Portland Place, London W1N 4BN, provide library facilities, open every weekday, to graduates in science and allied disciplines. The library is home to the Media Resource Service which offers journalists access to expertise on any scientific topic. Media relations are also strengthened by regular press conferences and book launches, and by articles prepared by the Foundation's Science Writer in Residence. The Foundation offers accommodation and meeting facilities to visiting scientists and their societies.

Information on all Foundation activities can be found at http://www.novartisfound.demon.co.uk

Novartis Foundation Symposium 215

IMMUNOLOGICAL TOLERANCE

1998

JOHN WILEY & SONS

Chichester · New York · Weinheim · Brisbane · Singapore · Toronto

Published in 1998 by John Wiley & Sons Ltd,
Baffins Lane, Chichester,
West Sussex PO19 1UD, England

National 01243 779777
International (+44) 1243 779777
e-mail (for orders and customer service enquiries): cs-books@wiley.co.uk
Visit our Home Page on http://www.wiley.co.uk
or http://www.wiley.com

Other Wiley Editorial Offices

John Wiley & Sons, Inc., 605 Third Avenue,
New York, NY 10158-0012, USA

Weinheim · Brisbane · Singapore · Toronto

Novartis Foundation Symposium 215
viii+240 pages, 29 figures, 17 tables

Library of Congress Cataloging-in-Publication Data

Immunological tolerance.
p. cm. – (Novartis Foundation symposium ; 215)
Symposium on Immunological Tolerance, held at the Novartis Foundation, London, July 8–10, 1997 – Contents p.
Editors: Gregory R. Bock (organizer) and Jamie A. Goode.
Includes bibliographical references and indexes.
ISBN 0-471-97843-4 (hbk : alk, paper)
1. Immunological tolerance – Congresses. 2. Autoimmunity – Congresses. I. Bock, Gregory. II. Goode, Jamie. III. Symposium on Immunological Tolerance (1997 : Novartis Foundation) IV. Series.
[DNLM: 1. Immune Tolerance congresses. 2. Autoimmune Diseases – immunology congresses. QW 504 I358 1998]
QR188.4.I428 1998
616.07′9–dc21
DNLM/DLC
for Library of Congress 98–14471
CIP

British Library Cataloguing in Publication Data

A catalogue record for this book is available from the British Library

ISBN 0 471 97843 4

Typeset in 10½ on 12½ pt Garamond by Dobbie Typesetting Limited, Tavistock, Devon.
Printed and bound in Great Britain by Biddles Ltd, Guildford and King's Lynn.
This book is printed on acid-free paper responsibly manufactured from sustainable forestry, in which at least two trees are planted for each one used for paper production.

Contents

Symposium on Immunological tolerance, held at the Novartis Foundation, London, 8–10 July 1997

Editors: Gregory R. Bock (Organizer) and Jamie A. Goode

This symposium is based on a proposal made by Abul K. Abbas

Participants

A. K. Abbas Immunology Research Division, Department of Pathology, Brigham and Women's Hospital and Harvard Medical School, Boston, MA 02115, USA

P. M. Allen Center for Immunology and Departments of Pathology and Pediatrics, Washington University School of Medicine, 660 S. Euclid Avenue, St Louis, MO 63110, USA

J. P. Allison Cancer Research Laboratory, University of California, Berkeley, CA 94720-3200, USA

B. Arnold Division of Molecular Immunology, Tumor Immunology Program, German Cancer Research Center, Im Neuenheimer Feld 280, 69120 Heidelberg, Germany

R. J. Cornall Nuffield Department of Medicine, Oxford University, John Radcliffe Hospital, Headington, Oxford OX3 9DU, UK

M. E. Digan Novartis Research Institute, 59 Route #10, East Hanover, NJ 07936, USA

D. A. Hafler Center for Neurologic Diseases, Harvard Medical School, 77 Avenue Louis Pasteur, Brigham and Women's Hospital, Boston, MA 02115, USA

G. J. Hämmerling Division of Molecular Immunology, Tumor Immunology Program, German Cancer Research Center, Im Neuenheimer Feld 280, 69120 Heidelberg, Germany

J. I. Healy Department of Microbiology and Immunology, Howard Hughes Medical Institute, Beckman Center, Room B169, Stanford University School of Medicine, Stanford, CA 94305, USA

M. K. Jenkins Department of Microbiology, Center for Immunology, University of Minnesota Medical School, Box 334 FUMC, 420 Delaware Street SE, Minneapolis, MN 55455, USA

T. Kamradt Rheumatologie und Klinische Immunolgie, Charité Medizinische Universitätsklinik III, D-10098 Berlin, Germany

D. Kioussis Division of Molecular Immunology, National Institute for Medical Research, Mill Hill, London NW7 1AA, UK

C. Kurts The Walter and Eliza Hall Institute of Medical Research, Post Office, Royal Melbourne Hospital, Victoria 3050, Australia

R. Lechler Department of Immunology, Royal Postgraduate Medical School, Du Cane Road, London W12 0NN, UK

M. J. Lenardo Laboratory of Immunology, National Institute of Allergy and Infectious Diseases, National Institutes of Health, Bethesda, MD 20892, USA

D. Mason Sir William Dunn School of Pathology, University of Oxford, South Parks Road, Oxford OX1 3RE, UK

N. A. Mitchison Department of Immunology, University College London Medical School, Windeyer Building, Cleveland Street, London W1P 6DB, UK

S. C. Schneider La Jolla Institute for Allergy and Immunology, 10355 Science Center Drive, San Diego, CA 92121, USA

E. M. Shevach Laboratory of Immunology, National Institute of Allergy and Infectious Diseases, National Institutes of Health, Bethesda, MD 20892, USA

K. Simon Centre d'Immunologie Marseille-Luminy, Case 906, 13288 Marseille cedex 9, France

B. Stockinger Division of Molecular Immunology, National Institute for Medical Research, Mill Hill, London NW7 1AA, UK

H. Waldmann Sir William Dunn School of Pathology, South Parks Road, Oxford OX2 7QY, UK

D. C. Wraith Department of Pathology and Microbiology, University of Bristol, School of Medical Sciences, Bristol BS8 1TD, UK

Introduction

N. A. Mitchison

Department of Immunology, University College London Medical School, Windeyer Building, Cleveland Street, London W1P 6DB, UK

The need for self/non-self discrimination has long been realized. Right from the turn of the century, when Ehrlich first began to formulate ideas about how the immune system might operate, it was clear that the system would need to have machinery that would allow it to recognize and resist infection, but would prevent it from attacking normal components of the body. Modern thinking about the problem traces back to the ideas set out in Burnet and Fenner's monograph on the immune response, published in the early 1940s. There they speculated that self-components might bear some kind of flag, which the immune system would recognize as a signal not to attack. Although that proposal turned out to be mistaken, it served the important purpose of preparing the ground for the experimentalists who began to address the problem a few years later.

It is a curious fact that when the experimentalists did get to work, starting in the late 1940s, they all chose to study tissue transplantation of one sort or another. From a biological point of view that was entirely logical, since cells transplanted from a foreign but related donor were about as close to the borderline between self and non-self as could be managed. Thus Medawar in London studied tissue transplanted between different strains of mice, Owen in Wisconsin studied haemopoietic cells naturally transplanted between cattle twins, and Hasek in Prague studied chicken eggs joined together in parabiosis. It was this work which lead to the definition of 'immunological tolerance' and to the Nobel prize awarded to Burnet and Medawar. Viewed from a modern perspective the logic behind this choice of material is not compelling, since the differences between proteins as they occur in different members of the same species are not fundamentally different from those that occur over longer evolutionary distances. Nevertheless, the choice was extraordinarily lucky, since it focused attention right from the start on antigens recognized by T cells, even though it was not clear at the time whether antigens—as the term was then used by microbiologists—were involved at all, and it took at least another decade to recognize T cells as a distinct compartment of the immune system. Nevertheless, it is T cells which are now known to play the main part in distinguishing between self and non-self.

From this early era, running on say to 1960, three cornerstones abide. One is that the lymphocyte (or the T cell as we now know) is, in Simonsen's phrase, the 'immunologically competent cell'. T cells can on their own initiate an immunological response, as Simonsen demonstrated by dropping them in small numbers onto the outer vascularized membrane of the chicken egg or injecting them into one of the membrane's blood vessels. The second is that tolerance operates through some kind of loss or deletion among T cells, since the flag theory had been rejected and the other major alternative, suppression, largely negated by experiment. That left the field open for Burnet's second theoretical contribution, the clonal theory of deletion, which was then rapidly accepted. The third concerns what is now called 'peripheral tolerance': the induction of tolerance among T cells which have completed their development and migrated out of the thymus into the peripheral lymphoid organs. It was observed that mice exposed to foreign antigen in the form of soluble monomeric protein would under some circumstances not make a positive immune response, but would rather become immunologically tolerant. Indeed, when individuals are exposed to such proteins without adjuvants, tolerance appears as the normal outcome.

In that era the students of immunological tolerance had no doubt about the wider implications of their work. Their principle aim was to learn how the immune system becomes adapted to the body in which it functions. They knew that skin and other non-immunological parts of the body have antigens which are not represented in the immune system. The system needs somehow to learn not to attack the cells which bear these antigens, and they saw the acquisition of tolerance as the means of it doing so. They also had other aims of a more practical nature, particularly achieving tolerance of organ transplants. They dreamed that tolerance of cancer cells might be broken.

Before moving on to more recent developments, let me expand this account of peripheral tolerance. As originally studied, the critical point in inducing peripheral tolerance was to avoid provoking a positive response that would mask it. This could be achieved experimentally for a wide variety of proteins, in various ways depending on the propensity of the protein to immunize (its 'immunogenic' activity), and its survival time in the body. With foreign immunoglobulin G, as studied by Dresser and Weigle, a single injection of de-aggregated protein was sufficient to induce a high degree of tolerance, probably because the protein continues to circulate in the body for many days. With other proteins such as foreign serum albumin repeated injections were needed, especially when a low-level positive response occurred and the repeated injections were needed to maintain continued exposure to the antigen. With even more immunogenic antigens, sub-lethal irradiation could be used to damp down interference from the positive response. A remarkable consistency of dose requirement was observed over this wide range of proteins, which extended to ovalbumin, bacterial lysozyme and diphtheria toxoid, and has recently been further extended to hen egg lysozyme expressed off a transgene, and ovalbumin as tested on TCR-transgenic T cells. The threshold is at 10^{-8}–10^{-9} M concentration in the body. Remarkably, this is the same threshold as applies in central tolerance (induced in the thymus), as judged by experiments in which tolerance was maintained rather than newly induced, or induced in thymus cultures *in vitro*. Indeed, the only noticeable difference between central and peripheral tolerance lies in the kinetics: induction in the thymus is faster.

One more observation is relevant before a model can be constructed. This is that a mouse which would normally be susceptible to tolerance induction by foreign IgG can be rendered resistant by enhancing its capacity to make a positive response. This could be done by administering whole bacteria, lipopolysaccharide or interleukin 1. Taken all together, these findings suggest that the immune system can be regarded as divided into two compartments: one in which positive responses are initiated, and another where tolerance is induced. In the thymus, only the tolerizing compartment is represented. In the peripheral immune system both are present; the tolerizing compartment is relatively small, to account for the slow kinetics, and shrinks still further in response to danger signals. The two compartments are defined functionally rather than anatomically, by 'activation states' of dendritic cells and T cells. We are still woefully ignorant of what constitutes activation in this sense, although CTLA-4 and the other co-stimulatory molecules are clearly molecular components of activation/quiescence. This unitarian view of low-dose tolerance would collapse if the pathways leading to T cell death were found to differ in the thymus and the periphery, but neither the molecular studies described here by Simon nor the genetics by Abbas suggest this.

Let me cite, self-indulgently, three fairly inaccessible reviews of mine on the subject. One is a history (Mitchison 1996), another summarizes the model of peripheral tolerance presented above together with its supporting data (Mitchison 1993), and a third reviews the balance between triggering a positive response and inducing tolerance (Kamradt & Mitchison 1997). Tolerance induction in the thymus is described in a more accessible report which I cite with particular pleasure, as it confirms a quantitative prediction which would have been made 20 years earlier (Robertson et al 1992).

In my opinion the most important recent developments in the field have been in immunological neglect and the augmenting of tolerance by immunoregulation. Neglect is a term introduced in this context by Zinkernagel, originally to describe a condition observed in mice transgenic for a viral antigen under the control of a tissue-specific promoter. When the antigen is expressed only on islet cells, it entirely fails to provoke an immune response, although the same cells are destroyed by a response provoked by infection with live virus. In the same direction, Sercarz has observed that many of the potential peptide epitopes on protein antigens get 'lost in the maelstrom' (his term) of peptides that compete with one another within antigen-presenting cells (see Schneider's contribution to this volume). Evidently the immune system need be tolerant of far fewer epitopes than had originally been thought. To use another phrase, the 'peptidic self' turns out to be far smaller than had been imagined. The other side of this coin is the evident ability of the system to down-regulate autoimmune responses, as described in this volume by Shevach and Arnold, and mentioned by Mason. Again, there seems to be far less need of tolerance.

Finally, a comment concerning activation-induced cell death ('AICD' as defined here by Abbas) and anergy (as described here by Hämmerling and applicable also to Jenkins' transfer model). These are both fairly new concepts, whose importance in the normal working of the immune system remains uncertain. In particular, it is quite unclear whether they have much to do with peripheral tolerance, as described above. It is true that AICD and peripheral tolerance share apoptosis as a common feature, but so do many other biological phenomena (discussed here by Lenardo). My guess is that AICD will turn out to have more to do with homeostasis of the T cell compartment as a whole, and with termination of the normal immune response, than with peripheral tolerance in the older sense. As for anergy, there is no doubt that progress is slowly but steadily being made in characterizing its molecular basis; and there is no doubt also that Jenkins' system offers a promising way of assessing its biological importance. Nevertheless, that assessment has still to be made, and in the meantime a question mark hangs over the phenomenon. As has been remarked many times, if anergy really is an important component of self-tolerance, it should be possible to pull cells anergized by self-antigens out of the normal T cell repertoire. That has not yet been done.

Readers of this volume should not be disappointed that so little can be found here of evaluation of the importance of the various mechanisms in the overall working of the immune system. The fact is that the system turns out to be far, far more complex than the pioneers had imagined. The problem posed by Ehrlich nearly a century ago is still with us. What we now know is that it isn't going to have a simple solution.

References

Mitchison NA 1996 Immunological tolerance. Encyclopedia of human biology, second edition. Academic Press, London. 10:113–126

Mitchison NA 1993 T cell activation states. The Immunologist 1:78–80

Kamradt T, Mitchison A 1997 Triggering the immune response by proinflammatory cytokines. J Biosciences 22:1–7

Robertson K, Simon K, Schneider S, Timms E, Mitchison A 1992 Tolerance of self induced in thymus organ culture. Eur J Immunol 22:207–211

Mechanisms of peripheral T cell tolerance

Luk van Parijs, Victor L. Perez and Abul K. Abbas

Immunology Research Division, Department of Pathology, Brigham and Women's Hospital and Harvard Medical School, Boston, MA 02115, USA

Abstract. Peripheral tolerance to self proteins is induced because these antigens are presented to T lymphocytes under conditions that do not allow effective immune responses to develop, or because the responses of the specific T cells are tightly regulated. The two principal mechanisms of peripheral tolerance are activation-induced cell death (AICD) and anergy. In $CD4^+$ T lymphocytes, AICD is induced by repeated stimulation, with high levels of interleukin (IL)-2 production. Under these conditions, the T cells co-express Fas (CD95) and Fas ligand (FasL), and engagement of Fas triggers apoptotic death of the T cells. Mice with defects in Fas, FasL, IL-2 or the IL-2Rα or β chain exhibit defects in AICD and develop autoimmune disease. The induction of T cell anergy is dependent on the recognition of B7 co-stimulators by the inhibitory T cell counter-receptor, CTLA-4. Failure of anergy is the likely basis for the fatal autoimmune disease of CTLA-4 knockout mice. The single-gene defects that result in autoimmunity are all defects in lymphocyte regulation, indicating that tolerance is often maintained by the control of lymphocyte responses to self antigen. The existence of distinct pathways of T cell tolerance suggests that different types of self antigens induce tolerance by distinct mechanisms.

1998 Immunological tolerance. Wiley, Chichester (Novartis Foundation Symposium 215) p 5–20

Immunological tolerance, or unresponsiveness, to self antigens is induced either by the encounter of immature lymphocytes with self antigens in the generative lymphoid organs, i.e. the bone marrow and thymus (central tolerance), or by exposure of mature lymphocytes to self antigens under particular conditions in peripheral tissues (peripheral tolerance). Peripheral tolerance is the mechanism which maintains unresponsiveness to antigens that are present only in peripheral tissues and not in the generative lymphoid organs. Peripheral mechanisms may also inactivate or kill lymphocytes that are specific for ubiquitous self antigens but escape central tolerance, for any reason. Recently developed experimental models are providing valuable information about the mechanisms of peripheral tolerance, and the pathophysiological consequences of failure of these mechanisms. This chapter summarizes our studies on peripheral T cell tolerance using a variety of transgenic and gene knockout mouse models.

Tolerance in $CD4^+$ helper T lymphocytes may be induced either by a process of functional inactivation (anergy), or by apoptotic cell death as a result of antigen stimulation (Fig. 1). It is also possible that the functions of $CD4^+$ T cells are suppressed by other cells, but there is little evidence that self antigens normally induce suppressive lymphocytes. The consequence of antigen recognition, i.e. activation or tolerance, depends mainly on two factors: how the antigen is presented to lymphocytes (its concentration, tissue location and persistence, and the nature of the cells that present the antigen), and how the responses of specific lymphocytes to that antigen are regulated. The importance of lymphocyte regulatory mechanisms is highlighted by recent findings, discussed below, that the disruption of several genes that control T cell responses leads to a failure of self-tolerance and the development of autoimmune diseases. One implication of these results is that recognition of self antigens in peripheral lymphoid tissues is a normal phenomenon, but pathological autoimmunity is prevented by controlling what happens to lymphocytes after they recognize and respond to self antigens. In the remainder of this chapter we will describe our analysis of two major pathways of peripheral T cell tolerance.

Activation-induced cell death

Activation-induced cell death (AICD) is a process of apoptosis induced by repeated activation of T lymphocytes by their cognate antigen. In $CD4^+$ T cells the principal mechanism of AICD is the co-expression of Fas (CD95) and Fas ligand (FasL), followed by engagement of Fas and delivery of a death-inducing signal (Nagata 1994). The importance of this mechanism for the maintenance of self-tolerance is most dramatically illustrated by the fatal lupus-like systemic autoimmune disease that develops in mice homozygous for mutations in either *fas* or *fasL*, and by the similar disease seen in humans with mutations in *fas* (Lenardo 1996).

Role of Fas in peripheral T cell tolerance

The idea that Fas–FasL interactions are involved in peripheral, but not central, T cell tolerance was first suggested by studies with superantigens in mice. This idea was convincingly established by our experiments using mice in which a transgenic T cell receptor (TCR) specific for a known peptide was bred onto the Fas-deficient *lpr*/*lpr* background (Singer & Abbas 1994). When these mice were treated with the cognate peptide antigen, the Fas deficiency caused a failure of peripheral T cell deletion but not deletion of specific T cells in the thymus. However, it is possible that some thymocyte populations are deleted by a Fas-dependent mechanism

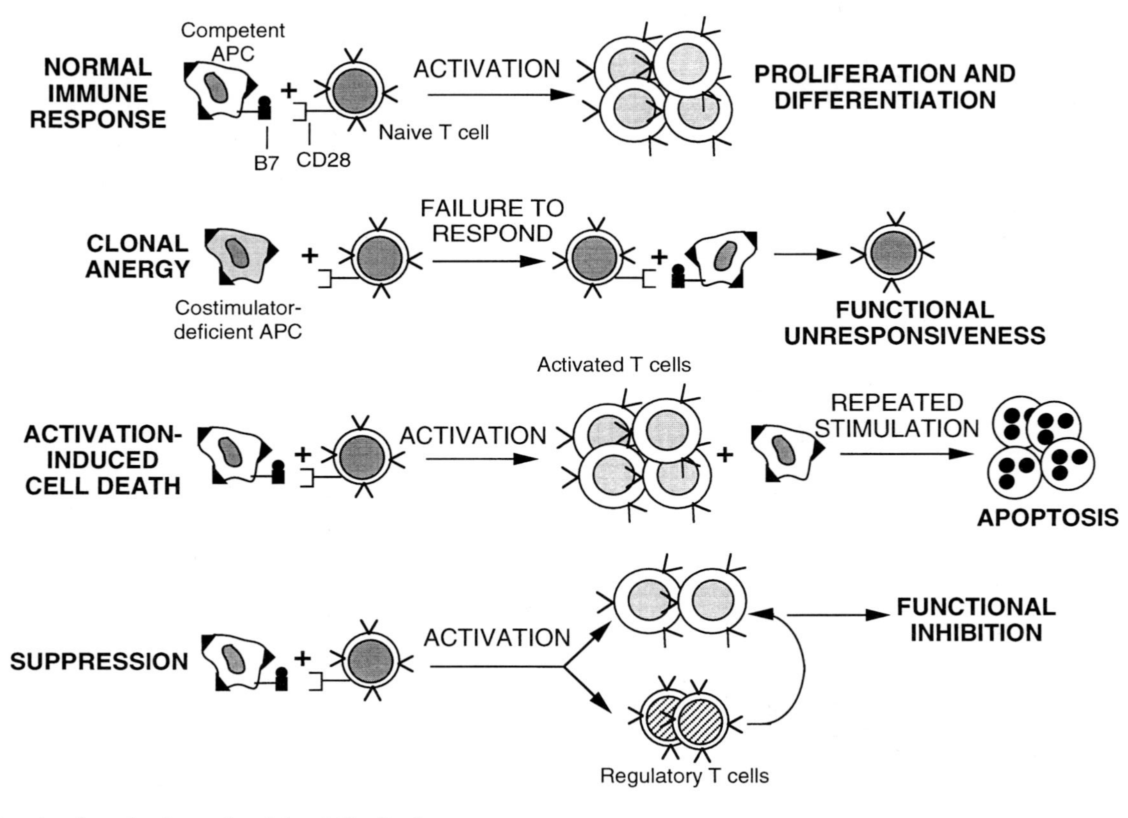

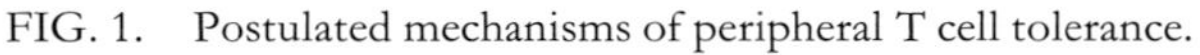

FIG. 1. Postulated mechanisms of peripheral T cell tolerance.

(Castro et al 1996, Kishimoto & Sprent 1997). Moreover, injections of high doses of peptide antigens may not be an appropriate way of mimicking the continuous presence of self antigens *in vivo*. In order to model self antigen exposure better, we have bred a hen egg lysozyme (HEL)-specific TCR, called 3A9, into wild-type or *lpr/lpr* H-2^k mice, and crossed them with H-2^k mice expressing transgene-encoded HEL as a 'self' antigen, also on the wild-type or *lpr/lpr* background (van Parijs et al 1998). Exposure to this self antigen leads to comparable thymic deletion of specific T cells in wild-type and *lpr/lpr* animals. We, therefore, conclude that Fas does not play a significant role in central tolerance, but is critical for deletion of mature T cells. It has also been suggested that tumour necrosis factor (TNF), which is homologous to FasL, may participate in AICD in mature $CD4^+$ T cells (Sytwu et al 1996). We have not seen a role for TNF in two TCR transgenic mouse lines, and mice lacking TNF or TNF receptors do not develop autoimmunity. Therefore, the major pathway of AICD-dependent self-tolerance in mature $CD4^+$ T cells appears to involve Fas–FasL interactions.

AICD and passive cell death

Two other important features of AICD have emerged from our recent studies. First, Fas-mediated AICD is distinct from a pathway of apoptosis, which we call passive cell death, that is a result of inadequate lymphocyte stimulation or depletion of growth factors (Fig. 2). Passive cell death is not Fas-mediated, and is prevented by the maintenance of high levels of Bcl-2 in T cells. In contrast, Bcl-2 does not block Fas-mediated AICD in T cells; whether or not other members of the Bcl family will interfere with AICD remains an open question. The main physiological role of Fas-mediated AICD is to eliminate T cells that are repeatedly stimulated by high concentrations of persistent antigens, e.g. self antigens. Passive cell death regulated by Bcl-2 functions mainly to promote lymphocyte survival either in the absence of cognate antigen (van Parijs et al 1998), or as levels of antigen decline during immune responses. Thus, these two pathways of apoptosis serve quite distinct functions in the immune system.

Stimuli that regulate AICD and passive cell death

Studies aimed at analysing the stimuli, other than antigen, that regulate AICD have revealed some interesting and surprising results. First, it has become clear that co-stimulation via CD28, which promotes T cell survival and responses to antigens, does not prevent Fas-mediated AICD (Fig. 3). In fact, the major effects of co-stimulation may be to promote the expression of survival genes such as *bcl-x*$_L$, and growth factors such as interleukin (IL)-2, both of which act to prevent passive cell death (van Parijs et al 1996). Second, IL-2, considered to be the

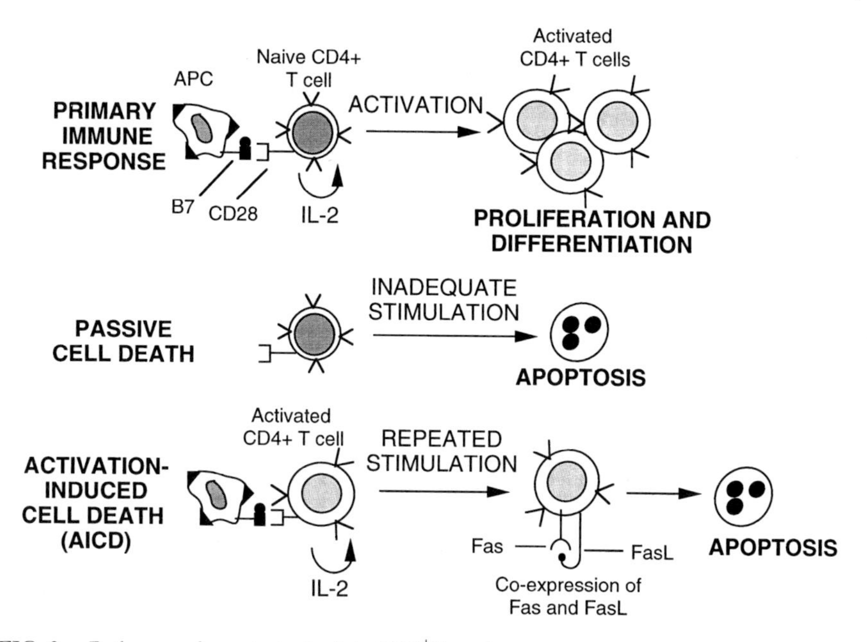

FIG. 2. Pathways of apoptotic death in $CD4^+$ lymphocytes.

prototypical T cell growth and survival factor, is also an important potentiator of Fas-dependent AICD (Fig. 3). This was first demonstrated by Lenardo (1991), and the obligatory role of IL-2-mediated signals in AICD has been convincingly established by the finding that T cells from TCR transgenic mice lacking the IL-2 receptor α chain, CD25, are resistant to Fas-mediated AICD (van Parijs et al 1997a). Failure of this IL-2-potentiated death pathway may account for the autoimmune disease seen in knockout mice lacking IL-2 or the α or β chain of the IL-2 receptor (Sadlack et al 1993, Suzuki et al 1995, Willerford et al 1995). Interestingly, one of the gene loci linked to type I diabetes in non-obese diabetic (NOD) mice maps to a site that also contains the *IL-2* gene (Denny et al 1997). The mechanism by which IL-2 potentiates Fas-mediated apoptosis is not well-defined. It is known to enhance the expression of FasL, and IL-2 may also promote the association of various proteins with the cytoplasmic domain of Fas to constitute a functional death complex (Muzio et al 1996). It appears that no other cytokine is as effective as IL-2 in potentiating AICD. Moreover, IL-2, like CD28-mediated co-stimulation, clearly prevents passive cell death, mainly by increasing expression of Bcl family members (van Parijs et al 1997a; Fig. 3). The differential effects of these stimuli on passive cell death and AICD further emphasize the fundamental biological differences between these two pathways of apoptosis.

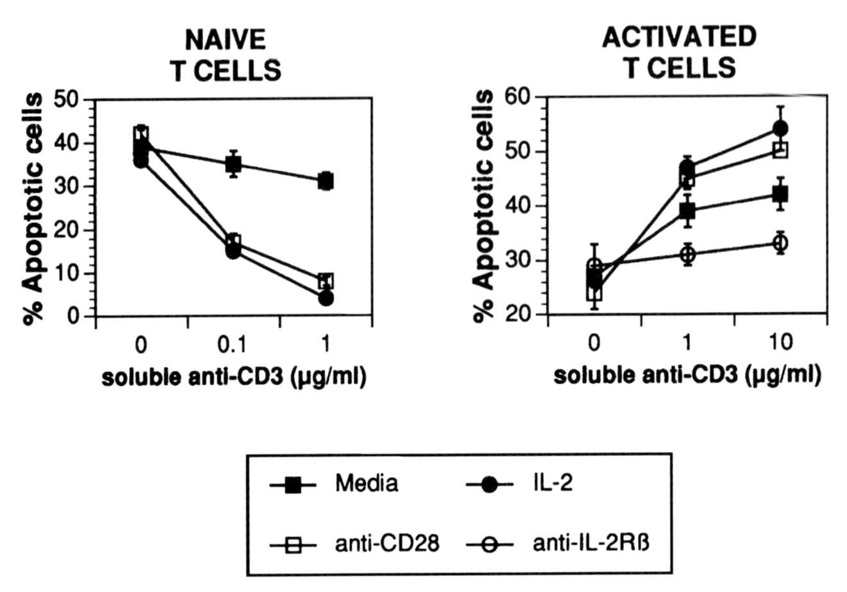

FIG. 3. Regulation of passive and active T cell apoptosis by CD28 and IL-2. Naïve $CD4^+$ T cells were purified from the spleens of transgenic mice. Some of these cells were activated *in vitro* with cognate peptide and syngeneic antigen-presenting cells, as described in van Parijs et al (1996). T cells were cultured with increasing concentrations of an anti-CD3 antibody, with and without IL-2 (50 units/ml), anti-CD28 antibody (1 μg/ml) or anti-IL-2Rβ antibody (10 μg/ml). After 24 h (activated T cells) or 48 h (naïve T cells), the percentages of apoptotic cells were assayed by staining with propidium iodide and flow cytometry (van Parijs et al 1996). Note that anti-CD28 and IL-2 protect naïve T cells from passive death, but potentiate SCID in previously activated T cell populations.

Anergy

The second mechanism of peripheral T cell tolerance, other than AICD, is anergy. This was first described in mouse T cell clones, and shown to be due to a block in antigen receptor-generated signals as a result of antigen recognition in the absence of co-stimulation and IL-2 (Schwartz 1990, 1996). Such results established the notion that functional responses to antigen (signal 1) require a second signals(s), provided by co-stimulators and/or growth factors, and signal 1 alone leads to functional anergy. Our initial attempts to define the importance of anergy in the induction of tolerance in normal T cells were done *in vitro*. These experiments were not informative, because if T cells from TCR transgenic mice are exposed to their cognate peptide antigen in the absence of co-stimulation or growth factors, they undergo passive cell death (van Parijs et al 1996, 1998).

We therefore focused our attention on an *in vivo* model of T cell anergy, first described by Kearney et al (1994), in which T cells expressing a transgenic TCR are 'parked' in normal syngeneic recipients and exposed to cognate antigens in an immunogenic or tolerogenic form. Exposure to tolerogenic peptide (high doses without adjuvants) leads to an initial expansion of specific T cells, but the T cells become unresponsive to subsequent re-stimulation with antigen *in vivo* or *in vitro* (Perez et al 1997). Moreover, even the tolerant T cells express some markers of activation. These results suggested to us that tolerance may be an abortive T cell response. Since all known responses of naïve T cells are dependent on co-stimulation, we asked if the same was true of tolerance induction. In fact, if co-stimulators of the B7 family are blocked at the time tolerogenic antigen is administered, the specific T cells do not become tolerant, but remain in a naïve yet functionally competent state. This result led to the surprising conclusion that *in vivo*, the induction of anergy may require co-stimulation, rather than being due to the absence of co-stimulation, as had been believed. The discovery of the inhibitory T cell receptor for B7 molecules, i.e. CTLA-4, provided an obvious candidate for a tolerance-inducing protein. In fact, blocking CTLA-4 completely prevents tolerance induction; in contrast, blocking CD28 prevents T cell priming by immunogenic forms of antigen (Perez et al 1997). These results suggest that co-stimulators of the B7 family may play dual functions in T cell responses to antigen (discussed below). More recently we have found that full T cell responses *in vivo* require co-stimulation and inflammatory cytokines, the prototype of which is IL-12. In fact, blocking CTLA-4 and administering IL-12 is sufficient to convert a tolerogenic stimulus (aqucous peptide antigen without antigen) to a fully immunogenic stimulus (van Parijs et al 1997b). The critical role of CTLA-4 in the induction of T cell anergy provides a mechanistic explanation for the fatal autoimmune disease that develops in CTLA-4 knockout mice (Tivol et al 1995, Waterhouse et al 1995).

Functions of co-stimulators in T cell activation and anergy

On the basis of the results summarized above, we conclude that co-stimulators of the B7 family play critical roles in regulating the choices between T cell survival, proliferation and differentiation on the one hand, and anergy or apoptosis on the other. These choices are not determined simply by the presence or absence of co-stimulation. In fact, our results show that if antigen is presented by antigen-presenting cells (APCs) in the presence of B7 antagonists, the antigen is effectively ignored by specific T cells (Fig. 4). This results in passive cell death *in vitro* and no response *in vivo* (the *in vivo* survival of non-responding T cells being uncertain at present). If, however, T cells use the CTLA-4 receptor to interact with B7 molecules at the time of antigen recognition, the result is T cell anergy,

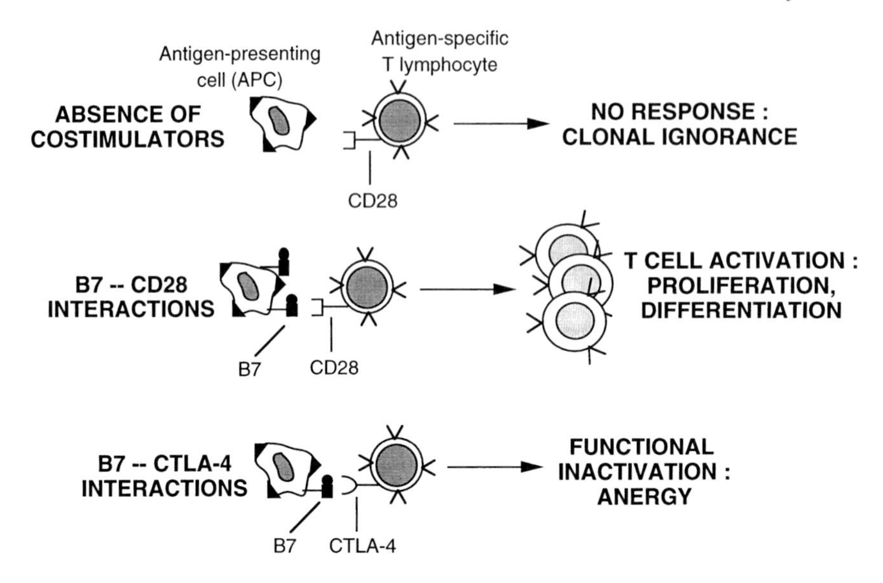

FIG. 4. A model for the role of co-stimulation in T cell activation and anergy.

whereas CD28–B7 interactions trigger functional T cell activation. What determines the choice between CTLA-4 vs. CD28 recognition of B7 is not established. One possibility is that, since CTLA-4 binds to B7 molecules with much higher affinity than does CD28 (Linsley et al 1991), low levels of B7 expression may preferentially engage CTLA-4 rather than CD28. However, this choice may also be regulated by other stimuli present at the time of antigen recognition, such as inflammatory cytokines.

Conclusions

The elucidation of T cell death pathways and mechanisms of anergy has provided important clues about how peripheral tolerance to self antigens is maintained. These concepts have been strongly reinforced by the identification of genetic mutations that lead to autoimmunity. Thus, the available evidence indicates that mutations in Fas/FasL and IL-2/IL-2Rα and β interfere with AICD, and this may lead to autoimmunity. In contrast, the disease of CTLA-4 knockout mice may be due to a failure of T cell anergy. An interesting question that arises concerns why, in these single gene mutants, anergy does not compensate for the defect in AICD, and vice versa. One possible answer is that AICD is responsible for tolerance to abundant and widely disseminated protein antigens, which may be presented by competent APCs and trigger the Fas pathway. In contrast, tissue antigens may be presented by resting APCs in the absence of inflammatory cytokines, leading to CTLA-4-mediated anergy. If this hypothesis is correct, the target antigens of autoimmunity should be different in models in which failure of AICD or anergy results in autoimmune disease. Attempts to identify such self antigens is an important goal of research in autoimmunity.

References

Castro JE, Listman JA, Jacobson BA et al 1996 Fas modulation of apoptosis during negative selection of thymocytes. Immunity 5:617–627

Denny P, Lord CJ, Hill NJ et al 1997 Mapping of the IDDM locus Idd3 to a 0.35-cM interval containing the interleukin-2 gene. Diabetes 46:695–700

Kearney ER, Pape KA, Loh DY, Jenkins MK 1994 Visualization of peptide-specific T-cell immunity and peripheral tolerance induction *in vivo*. Immunity 1:327–339

Kishimoto H, Sprent J 1997 Negative selection in the thymus includes semimature T cells. J Exp Med 185:263–271

Lenardo MJ 1991 Interleukin-2 programs mouse $\alpha\beta$ T lymphocytes for apoptosis. Nature 353:858–861

Lenardo MJ 1996 Fas and the art of lymphocyte maintenance. J Exp Med 183:721–724

Linsley PS, Brady W, Urnes M, Grosmaire LS, Damle NK, Ledbetter JA 1991 CTLA-4 is a second receptor for the B cell activation antigen B7. J Exp Med 174:561–569

Muzio M, Chinnaiyan AM, Kischkel FC et al 1996 FLICE, a novel FADD-homologous ICE/CED-3-like protease. Cell 85:817–827

Nagata S 1994 Fas and Fas ligand: a death factor and its receptor. Adv Immunol 57:129–144

Perez VL, van Parijs L, Biuckians A, Zheng XX, Strom TB, Abbas AK 1997 Induction of peripheral T cell tolerance *in vivo* requires CTLA-4 engagement. Immunity 6:411–417

Sadlack B, Merz H, Schorle H, Schimpl A, Horak I 1993 Ulcerative colitis-like disease in mice with a disrupted interleukin-2 gene. Cell 75:253–261

Schwartz RH 1990 A cell culture model for T lymphocyte clonal anergy. Science 248:1349–1355

Schwartz RH 1996 Models of T cell anergy: is there a common mechanism? J Exp Med 184:1–8

Singer GG, Abbas AK 1994 The Fas antigen is involved in peripheral but not thymic deletion of T lymphocytes in T cell receptor transgenic mice. Immunity 1:365–371

Suzuki H, Kundig TM, Furlonger C et al 1995 Deregulated T cell activation and autoimmunity in mice lacking interleukin 2 receptor β chain. Science 268:1472–1476

Sytwu H-K, Liblau RS, McDevitt HO 1996 The roles of Fas/APO-1 (CD95) and TNF in antigen-induced programmed cell death in T cell receptor transgenic mice. Immunity 5:17–30

Tivol EA, Borriello F, Schweitzer AN, Lynch WP, Bluestone JA, Sharpe AH 1995 Loss of CTLA-4 leads to massive lymphoproliferation and fatal multi-organ tissue destruction, revealing a critical negative regulatory role of CTLA-4. Immunity 3:541–547

van Parijs L, Ibraghimov A, Abbas AK 1996 Role of co-stimulation and Fas in T cell apoptosis and peripheral tolerance. Immunity 4:321–328

van Parijs L, Biuckians A, Ibraghimov A, Alt FW, Willerford D, Abbas AK 1997a Functional responses and apoptosis in CD25 (IL-2Rα)-deficient lymphocytes expressing a transgenic antigen receptor. J Immunol 158:3738–3745

van Parijs L, Perez VL, Biuckians A, Maki RG, London GA, Abbas AK 1997b Role of IL-12 and costimulators in T cell anergy *in vivo*. J Exp Med 186:1119–1128

van Parijs L, Biuckians A, Abbas AK 1998 Functional roles of Fas and Bcl-2-regulated apoptosis of T lymphocytes. J Immunol 160:2065–2071
Waterhouse P, Penninger JM, Timms E et al 1995 Lymphoproliferative disorders with early lethality in mice deficient in CTLA-4. Science 270:985–988
Willerford DM, Chen J, Ferry JA, Davidson L, Ma A, Alt FW 1995 Interleukin-2 receptor alpha chain regulates the size and content of the peripheral lymphoid compartment. Immunity 3:521–529

DISCUSSION

Allison: Perhaps I could inject a note of caution about the interpretation of some of your experiments. Several years ago when many of us were beginning to work on the effect of CD28, we found that when we took anergized T cells and give them signal plus CD28 we could reverse anergy. The problem in the interpretation, of course, is that in these experiments it is not possible to track single cells. When we began to do single cell analysis to look at this phenomenon, we found that we had rescued a small fraction of cells that had not actually been anergized in the initial stimulation. Because we are giving vigorous and optimal co-stimulation these cells go nuts, which gives the appearance of reversing anergy. In fact, we had just selected out some of the cells.

In this context I want to inject a note of caution into the interpretation of your *in vivo* CTLA-4 blockade experiments. We have obtained similar data: we can apparently block the anergy that is induced by SEB (*Staphylococcus* enterotoxin B). The problem is that in your experiments it is possible that the same events are taking place: that is, covering up CTLA-4 allows a lot of cells to proliferate higher and so you see a response which might be interpreted as the reversal of anergy, but you haven't—you have just allowed a small population of cells to go nuts.

The biochemical data that we're developing suggest that CTLA-4 gets in the way of initial activation. It is difficult to put this in the context of anergy itself. Until you can track individual cells in some way, you won't know whether you are changing a physiological state of an individual cell or whether on a population basis you are just overexpanding a small number of cells. This is the real danger in interpreting these experiments.

Abbas: What you're saying is that blocking CTLA-4 is not preventing anergy, but it's compensating for the anergic state by allowing a very small population to dominate. I don't know if we could ever answer that. The question is whether that is a semantic difference as far as the physiology of the responses is concerned, or a real difference. Mechanistically, it's a real difference.

Some years ago, Beverly and co-workers had similar results on IL-2 reversing what was then called 'clonal anergy' (Beverly et al 1992). The problem has been that we have never been able to do those experiments with primary T cells from T cell receptor transgenics, because if we give them inadequate stimuli *in vitro*, the primary T cells die by neglect. So we just can't do those anergy experiments and re-visit the reversal phenomenon in primary T cells.

Mason: With regard to AICD, any activated T cell can suffer one of only two fates: either it dies or it turns into a memory cell. It may go through an effector phase in the meantime, which is the primary immune response, but apart from that, these are the only two fates. So when you talk about self-tolerance being induced by activation of T cells, how are you distinguishing self-tolerance from a physiological response to a foreign antigen?

Abbas: This is an important issue, because when cells die at the tail end of an immune response (and the majority do), that's not AICD: rather, it is death by neglect. If you over-express Bcl-2 you retard death by neglect. That does not happen in an *lpr* or a *gld* mouse, so if we take T cells expressing a particular TCR transgene on the *lpr* or *gld* mutation, adoptively transfer them into normal mice, immunize and then just follow them, they expand and decline like wild-type cells. So that death fate, which is the fate of the majority of T cells, is not activation induced, although it follows activation. What we refer to as AICD is the form of death that occurs with repeated activation, but presumably only happens with self-antigens, because they're persistent enough to give repeated activation. Most foreign antigens are eliminated.

Waldmann: For any abundant self-antigen, I would have thought that there will be central tolerance. Can you give me an example of abundant self-antigen that would have T cells available to it to produce AICD?

Abbas: I can't. On the basis of the specificity of the autoimmunity in the *lpr* and *gld* mouse, you would predict that antigens such as nucleosomal proteins and histones would be dealt with by AICD. But this is a sort of circular reasoning on the basis of what the autoimmunity is directed against in the mice with Fas defects. The question as to why central tolerance isn't good enough for dealing with histones and nuclear proteins is a good one. I don't know the answer.

Lenardo: People and animals that have Fas deficiencies have, for example, anti-platelet antibodies: I wouldn't expect the antigens on the surface of platelets to which these antibodies are directed to be seen in the thymus.

Abbas: You see haemoglobin in the thymus, so why shouldn't you see antigens expressed on the surface of a red cell in the thymus?

Lenardo: It hasn't been demonstrated.

Mitchison: Let me get one thing clear about the distinction you made between peripheral and central tolerance. You suggested that there were two different mechanisms involved. You said that AICD was the peripheral mechanism that didn't operate in the thymus, which is fair enough, but do you think that the thymic mechanism doesn't operate in the periphery?

Abbas: Negative selection in the thymus would also be antigen-induced and presumably would occur via apoptotic cell death. Physiologically it's the same, but what I was trying to emphasize is that there is no evidence that the Fas pathway plays any role there.

Kioussis: When we talk about negative selection in the thymus following peptide injections, we should be careful not to equate it with natural negative selection, because the former has two components. One is the recognition of the peptide, and the other is the activation of the peripheral T cells. Both contribute to the deletion of double-positive thymocytes, but it is difficult in experiments *in vivo* to distinguish between the two when you have peripheral activation and cognate recognition of the antigen by the double-positive thymocytes.

Abbas: But remember that we can take an HEL-specific T cell receptor transgene and cross it with a mouse that expresses HEL. In this situation there are no peripheral T cells, because they all get deleted in the thymus. This is as close as we can get to central tolerance to an endogenous self antigen. (This mouse is the high level HEL expressor with about 10 ng/ml serum HEL.) The whole reason for going through this cumbersome breeding was that we wanted to get away from the repeated peptide injection protocol. I believe that Chris Goodnow's lab has done pretty much the same experiment with *gld* mice and sees essentially the same thing — the HEL-specific T cells in the double transgenics all get deleted in the thymus.

Allison: A better way to do the experiment would be to use mixed chimeras.

Kioussis: That is the way we did it.

Abbas: And what was the result?

Kioussis: When we gave the peptide, the antigen-specific thymocytes disappeared. A lot of the non-transgenic T cells survived but their numbers were reduced.

Abbas: This is in *lpr* mice?

Kioussis: No, this is in wild-type.

Abbas: So it does not address the issue of Fas.

Jenkins: Abul, It seems to me that the problem is that for peripheral tolerance you have mature T cells who want to respond in a positive way and are programmed to do that, and yet they have to be able to stop short of this response if they are directed against a developmentally regulated self-antigen that is new on the scene. The simplest way for this to be achieved is still to activate the antigen-presenting cells in response to things from microbes and then induce co-stimulatory molecules that can then push the response to go beyond that abortive phase into full activation. How different is this new model? To get a response, you still need TCR plus enhanced expression of co-stimulatory molecules on APCs to get priming.

Abbas: That's really talking about functional inactivation. With the anergy model that we all talk about, whether it's based on CTLA-4 or whether it's simply lack of signal 1, you would still say the same thing: that once you trigger the antigen-presenting cell to express high levels of B7 and/or provide other second signals, you trigger a response. Saying that anergy requires CTLA-4/B7 interactions is not going to fundamentally change the way we've seen anergy in the past. It is going to make one big difference, in that attempts to induce anergy by blocking B7 are unlikely to be successful if anergy requires B7 recognition. If anergy requires CTLA-4/B7 interactions, then simply blocking B7 is not going to induce tolerance: it is going to induce failure to respond for as long as the antagonist hangs around. In fact, I think the emerging data support this idea. If you look at the transplant models in mice treated with CTLA-4 IgG, even though the graft isn't rejected, alloreactivity is maintained.

Mitchison: We're deluding ourselves if we lay too much emphasis on anergy. If you ask a representative sample of immunologists, most of them would express the opinion that anergy is merely some kind of *in vitro* phenomenon. Until somebody has a marker for anergy and can show that there are substantial proportion of anergized cells in the normal immune system, people just aren't going to listen to us.

Allison: I would second that. Your data can be explained by saying that what CTLA-4 does is to contribute to ignorance: it just keeps the cell from getting activated.

Abbas: No, it can't be ignorance: antigen has to recognized in order to trigger the CTLA-4/B7 recognition pathway. Therefore, by definition the cells are seeing something.

Allison: In the molecular sense, when you trigger CTLA-4 engagement you raise the threshold of T cell receptor signals that you need to progress to full activation. I can show you that when you engage CTLA-4 you don't get CD69, CD25 or IL-2. It's not that you get a response and then the cell becomes unresponsive: it basically shuts things off and raises the threshold.

Jenkins: You can have a model where you have to activate naïve T cells to a point where they can become susceptible to tolerance induction. Your data are completely consistent with this idea.

Abbas: I would not disagree with that. I think it is important to point out that the induction of Fas-mediated death is actually dependent on full activation. This is an important idea that just needs to be nailed down: it requires CD28 and it requires IL-2. That can't be due to something in the APC.

Lechler: I'm interested in the CTLA-4 data that you have. I wonder whether CTLA-4 simply limits the amount of co-stimulation, and therefore limits IL-2. As a consequence, the cells don't divide enough and, as Mark Jenkins showed a few years ago, a cell that doesn't divide enough ends up in the 'off' position. So the experiment I would be interested to see you do is, instead of blocking CTLA-4, after two days for you to block CD28 (using a Fab preparation of anti-CD28), which would limit co-stimulation through CD28. I suspect that you might end up with exactly the same effective tolerization, suggesting then that it's nothing specific about CTLA-4 signalling, which I think is what you are trying to argue.

Abbas: If we induce tolerance with either aqueous peptide or a large dose of peptide in incomplete Freund's adjuvant, and then block CD28, we don't get any more or less tolerance. It has no effect.

Lechler: In the experiment I propose you would have to allow the initial activation to occur before blocking CD28.

Abbas: We have not done that.

Stockinger: I think this CTLA-4 issue has caused some confusion. I don't think it has been proved that B7 interaction is necessary to induce anergy. CTLA-4 seems to me a nice pathway to get memory rather than full effector fate. Initially, you just brushed aside the issue of recognition in the absence of B7 as 'ignorance'. I would have thought that we can interpret ignorance as the fact that naïve T cells do not have access to certain areas where these APCs are found. Has it been ruled out that a cell can be partially activated if it sees class II peptide presented on the cell in the absence of B7?

Abbas: The B7 blocking data are in a paper that came out a couple of months ago (Perez et al 1997). If you induce tolerance and treat the mice with CTLA-4 Ig to block all B7 interactions, then the cells are phenotypically naïve (small in size, high CD45RB, low CD44). If you take them out they're functionally competent. It looks just like the old Ohashi/Zinkernagel experiments (Ohashi et al 1991). They respond normally to peptide *in vitro*; their response is that of a naïve population. If you leave them in and prime the animals a week or two later, they expand just as naïve cells do. That was the reason I said that if you block B7 you prevent anything from happening. The cells behave just as if they have not seen antigen. That is why I equated that with clonal ignorance.

Parenthetically I would add that in the CTLA-4 knockout mouse, treatment with CTLA-4 Ig blocks B7 and the mouse is normal. If you stop the B7 antagonist the mouse dies three or four weeks later of autoimmunity (Tivol et al 1997).

Hafler: I would like to explore further this issue of clonal ignorance and what it really means. The point that Paul Allen made many years ago was that we live in a sea of cross-reactivity. When we look at our T cell clones, it's hard not to find cross-reactivity to many different peptides. We're continually being exposed to cross-reacting peptides. This means there's a constant baseline stimulus of low-level signal going perhaps to all T cells without strong co-stimulatory signals. Are we inducing some type of not just clonal ignorance, but some sort of a negative influence? For example, when we stimulate T cells *ex vivo* with CHO cells transfected with DR2 plus myelin basic protein peptide by itself, we see a significant number of T cell clones that secrete low amounts of IL-4. When we add co-stimulatory signals, we drive them to clonal expansion. Is clonal ignorance just clonal ignorance, or is it a tonic stimulation as occurs in the nervous system driving it to secrete cytokines such as IL-4 to dampen the general immune response?

Abbas: I can't answer that. Much as I'm fond of the regulatory function of anti-inflammatory T cells, is there any evidence that self-antigens normally induce the development of Th2 cells? As far as I can tell there is no evidence that normal self-antigens will normally elicit Th2 responses.

Shevach: One model by which T cells could escape clonal deletion in the thymus is for them to become Th2 cells.

Abbas: Sure. That's why we all show it as anergy, deletion and suppression, but what is the evidence that we all walk around with self-reactive Th2 populations? The answer is, not much. We can't answer the question as to whether we are constantly being exposed to self-antigens and either ignoring the self-antigens or being pushed towards an anti-inflammatory pathway, which is really what you're asking.

Waldmann: The trouble with your question is that for peripheral self-antigens which are present in small amounts, if there was Th2 activity present, the level would be so low that you would not be able to measure it. With foreign antigens one can force the system.

Abbas: I'm not saying that it doesn't happen, I'm just saying that we don't have any solid information.

Mitchison: This is a basic question. As far as the prevention of autoimmune disease is concerned, you want to have a potentially protective population that is inducible. It doesn't matter whether it is there normally or not.

Shevach: As Jim Allison brought up earlicr, one mechanism for rescue of the response is that the responding T cell population is heterogeneous. You have an obvious heterogeneity in your system, in that some of the cells express endogenous receptor even in the transgenic. Are you worried by that?

Abbas: This is because of the lack of α-chain allelic exclusion. We have only recently got enough DO11.10 SCID mice to do the experiment. However, it is still not going to answer the heterogeneity question, because Jim Allison's question is not about endogenous antigen recognition, it's just the possibility that if 10 cells are tolerized and one is not, anti-CTLA-4 allows that one to proliferate and it dominates the response. We cannot prove or disprove this in the absence of a marker for the anergic cell.

Jenkins: DO11.10 SCID T cells can be tolerized by soluble antigen injection, so T cells expressing endogenous TCRs are not required.

References

Beverly B, Kang SM, Lenardo MJ, Schwartz RH 1992 Reversal of *in vitro* T cell clonal anergy by IL-2 stimulation. Int Immunol 4:661–671

Ohashi PS, Oehen S, Buerki K et al 1991 Ablation of 'tolerance' and induction of diabetes by virus infection in viral antigen transgenic mice. Cell 65:305–317

Perez VL, van Parijs L, Biuckians A, Zeng XX, Strom TB, Abbas AK 1997 Induction of peripheral T cell tolerance *in vivo* requires CTLA-4 engagement. Immunity 6:411–417

Tivol EA, Boyd SD, McKeon S et al 1997 CTLA-4Ig prevents lymphoproliferation and multiorgan tissue destruction in CTLA4-deficient mice. J Immunol 158:5091–5094

B cell antigen receptor signalling in the balance of tolerance and immunity

Richard J. Cornall and *Christopher C. Goodnow

*Nuffield Department of Medicine, Oxford University, John Radcliffe Hospital, Headington, Oxford OX3 9DU, UK and *Australian Cancer Research Foundation Genetics Laboratory, Medical Genome Centre, John Curtin School of Medical Research, The Australian National University, Mills Rd, PO Box 334, Canberra, ACT 2601, Australia*

Abstract. The quantity and quality of signals from the B cell antigen receptor (BCR) drives the positive and negative selection of B lymphocytes and establishes the balance of tolerance and immunity. Experiments using immunoglobulin transgenic mice and mutations in key BCR signalling components have given insight into how the antigen receptor is tuned and how thresholds for qualitatively different outcomes are established and maintained. This research also describes how genetic variants can shift the balance between autoimmunity and tolerance.

1998 Immunological tolerance. Wiley, Chichester (Novartis Foundation Symposium 215) p 21–32

The fate of B and T lymphocytes is very largely decided by signals which they receive through their surface antigen receptors (Goodnow 1996). Receptor signalling directs the positive and negative selection steps which shape the preimmune repertoire as well as the clonal expansion and production of antibody which follows infection. Each of these selection steps has its own activation threshold which has evolved as a compromise between the need to avoid harmful autoimmune disease and the primary role of the immune system which is to be able to fight a wide range of unpredictable infections. Understanding the biochemical and genetic basis of antigen receptor signalling thresholds is therefore of great importance in understanding how the balance between immunity and tolerance is maintained in normal and pathological states.

The structure and function of the B cell antigen receptor (BCR) is now relatively well understood (DeFranco 1987, Kurosaki 1997). The core of the BCR is composed of the transmembrane immunoglobulin receptor and associated molecules CD79α (Igα) and CD79β (Igβ). Antigen–receptor cross-linking leads to activation of Src-family kinases, Lyn, Lck, Blk, Fyn and Fgr, and phosphorylation of tyrosines within ITAM motifs in the CD79 cytoplasmic domains. The protein tyrosine kinase Syk associates with the phosphorylated CD79 ITAMs and is activated by phosphorylation. The importance and activation sequence of the various tyrosine kinases has not been unambiguously determined, nor has the issue of individual kinases, particularly Src kinases, which may have both unique and redundant functions. The concerted action of Syk, Src-family kinases and other downstream kinases, including Btk, initiates a cascade of biochemical pathways which generate second messengers, including the activation of the phosphatidylinositol pathway, the activation of Ras, the activation of the lipid-dependent kinases and the mobilization of calcium. In turn, the second messengers induce gene expression by translocating cytoplasmic transcription factors such as NF-AT and NF-κB to the nucleus and by activating mitogen-activated protein kinases (MAPKs) that phosphorylate existing nuclear transcription factors (Healy et al 1997). Our present understanding of the biochemistry of antigen receptor signalling provides the basis on which to explore the genetics and cellular biology of B cell selection.

The role of BCR signalling in selection of immature B cells

Experiments with immunoglobulin receptor transgenic mouse models and mutations in key BCR signalling molecules have revealed how negative signalling determines the fate of immature B cells which encounter self-antigen. There exists a spectrum of response which is dependent upon the strength of antigen receptor signalling (Goodnow et al 1995, and Fig. 1A). Multivalent membrane-bound self antigens, a covalent dimer of hen egg lysozyme or double-stranded DNA cause self-reactive B cells to undergo maturational arrest and elimination (Chen et al 1995, Hartley et al 1991, 1993, Nemazee & Bürki 1989, S. Akkaraju, J. Healy & C. C. Goodnow, unpublished results), either through cell death or by editing light chain genes to generate a new binding specificity. By this means it is possible to ensure highly robust tolerance to the most dangerous forms

---

FIG. 1. (*opposite*) Antigen receptor signalling determines the fate of immature and mature B cells. (A) Immature B cells are eliminated if they fail to generate a functional IgM receptor or if they bind multivalent antigen (i and iv). Between these two extremes of IgM-mediated signalling are naïve cells (ii), which fail to bind antigen with high affinity but have a functional receptor, and anergic cells (iii), which bind self-antigen of low avidity. (B) Anergic and naïve cells emerge form the bone marrow and mature to express both IgM and IgD. Naïve cells recirculate through the secondary lymphoid organs for about 4 weeks. Anergic cells are unable to compete with naïve cells for access to the B cell follicles and are trapped in the T cell zones, where, in the absence of T cell help, they die after about 3 days. (C) Naïve B cells which bind foreign antigen during their circulation through the secondary lymphoid organs also localize to the T cell zones and are excluded from the follicles. However, newly-stimulated mature B cells are rescued from apoptosis by T cell help, which leads to proliferation, differentiation into plasma cells and antibody production.

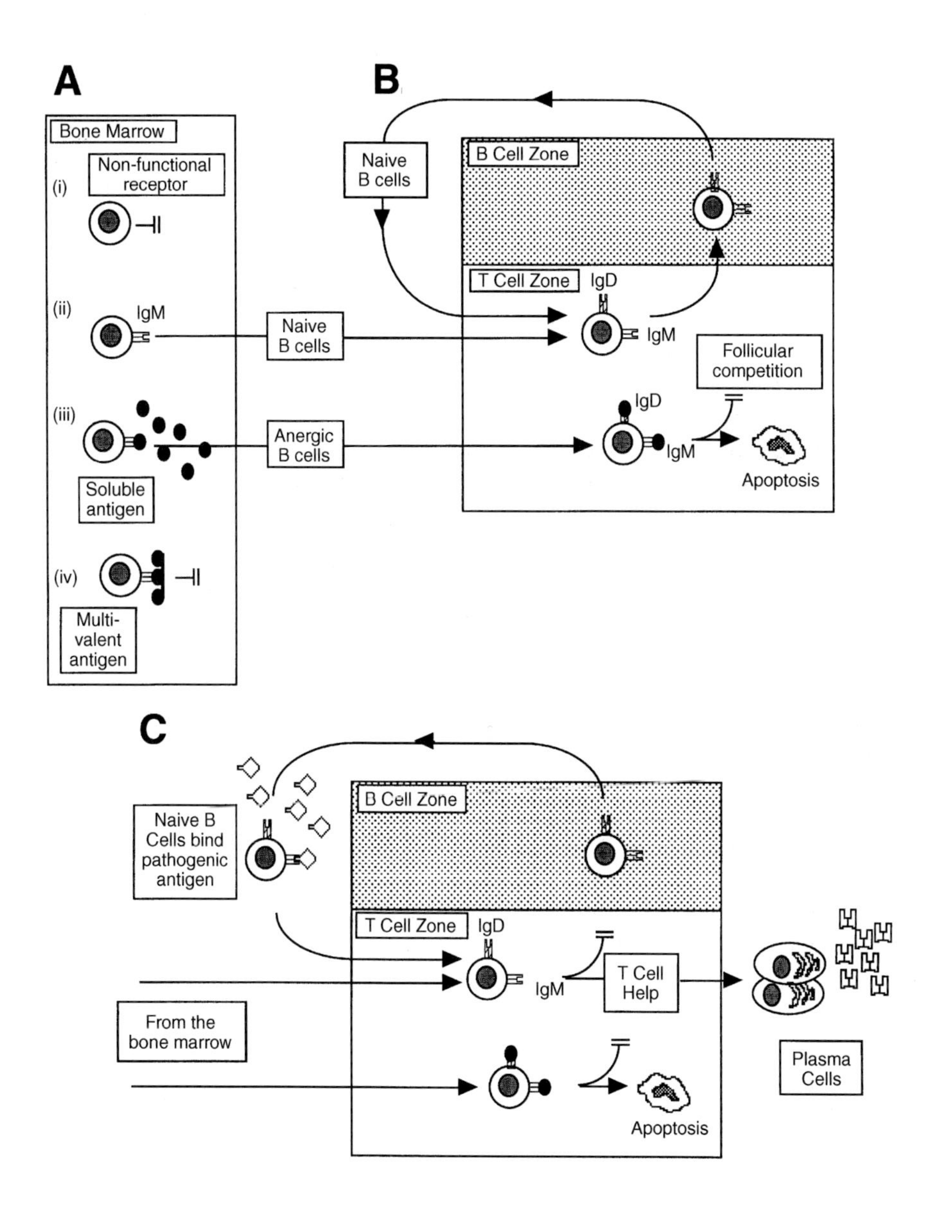
A
Bone Marrow
(i)
Non-functional receptor
(ii)
IgM
(iii)
Soluble antigen
(iv)
Multi-valent antigen
Naive B cells
Anergic B cells
B
Naive B cells
B Cell Zone
T Cell Zone
IgD
IgM
Follicular competition
IgD
IgM
Apoptosis
C
Naive B Cells bind pathogenic antigen
B Cell Zone
T Cell Zone
IgD
IgM
T Cell Help
From the bone marrow
Apoptosis
Plasma Cells

of self antigens, such as the ABO blood groups. Lower valency antigens, such as a soluble monomeric form of hen egg lysozyme (sHEL) or single-stranded DNA, generate a less stringent form of tolerance known as anergy (Goodnow et al 1988). Anergy is characterized by active desensitization of the BCR both by a profound block in proximal antigen receptor signalling (Cooke et al 1994) and a 10–50-fold modulation of surface immunoglobulin M class (IgM) receptors (Bell & Goodnow 1994). Anergic B cells express a unique profile of activated second messengers and transcription factors which are selectively dependent upon a low level spiking calcium flux (Healy et al 1997). Anergy is a process which compromises tolerance and immunity because anergic cells do leave the bone marrow, do mature, and despite a rapid turnover are available for activation by highly multivalent pathogenic antigens. When mechanisms for desensitizing the BCR are defective, as in the motheaten viable mutation in the cytosolic tyrosine phosphatase SHP-1, exaggerated BCR signalling to antigen causes deletion of B cells that would normally be rendered anergic (Cyster & Goodnow 1995a). Finally, at the far end of the spectrum are self-antigens which are at too low a concentration or too low an affinity to cause either inactivation or elimination of self-reactive cells (Adelstein et al 1991, Akkaraju et al 1997, Goodnow et al 1989). The B cells in these situations are functionally ignorant of the antigens, although they may show subtle modulation of their antigen receptors in response to chronic signalling. Many weakly self-reactive B cells may fall into this last category, in which the danger of autoimmunity is presumably strongly outweighed by potential to fight infections.

Signalling from the BCR is also a prerequisite for B cell survival at the immature cell stage and this ensures that only cells which are immunocompetent enter the preimmune repertoire. Mature B cells fail to accumulate in mice lacking positive-acting components such as the transmembrane segment of IgM heavy chain, the Igβ protein, the ITAM of Igα, Syk or Btk kinases, or CD45. In the absence of CD45, for example, B cells have a depressed intracellular calcium, ERK, and mitogenic response to antigen (Cyster et al 1996, Neel 1997). The depressed BCR signalling in the absence of CD45 remains sufficient for the elimination of self-reactive B cells by high avidity membrane-bound antigen, but is insufficient to promote survival of mature B cells in the absence of antigen, and self-reactive B cells that would normally become anergic to low avidity self antigen are positively selected (Cyster et al 1996).

The role of BCR signalling in the selection of mature B cells

Immunocompetent but naïve transgenic B cells which have yet to encounter antigen, emerge from the bone marrow and recirculate through the T and B cell zones and tissues of the secondary lymphoid organs during a lifespan of about four weeks (Fig. 1B). If on their travels these cells are stimulated by a pathogenic antigen, then they rapidly localize to the T cell zones of the spleen and lymph nodes, where, provided they can collaborate with T cell help, they proliferate and start to differentiate into plasma cells (Fig. 1C).

The same mechanism that causes naïve cells to localize to the T cell zones also causes the localization of anergic cells, provided that they exist within a polyclonal repertoire (Cyster & Goodnow 1995b, Cyster et al 1994). This process has been called follicular competition because it is dependent upon the presence of unstimulated B cells out-competing other cells for access to as yet undefined survival factors associated with the B cell follicles. After follicular exclusion, T cell help is required to rescue both naïve cells and anergic cells from apoptosis. The state of anergy itself and T cell tolerance to self-antigen ensure that anergic B cells are unlikely to encounter T cell help and under these circumstances they usually die within about three days (Fig. 1B). Competent T cells which do interact with anergic B cells, for example by cross-reacting epitopes, usually kill rather than activate the anergic cells by transmitting a Fas-dependent cell death signal (Rathmell et al 1995) (Fig. 2). Exclusion from follicles and rapid turnover in the outer T cell zones also appears to account for the failure of most new B cells to accumulate after emigrating from the bone arrow in adult non-transgenic animals (Chan & MacLennan 1993, Lortan et al 1987).

The polarized fates of the anergic and naïve immunocompetent cells seems to be determined by differences in BCR signalling as well as the state of T cell tolerance. The inability of the anergic B cells to effectively up-regulate B7.2 expression in response to antigen binding may be of particular importance, because the expression of a B7.2 transgene can rescue anergic B cells from the Fas-dependent cell death (J. C. Rathmell, S. Fournier, J. P. Allison and C. C. Goodnow, unpublished data). Under normal circumstances anergic cells in the T cell zones require binding to highly multivalent pathogens with cross-reacting TCR epitopes to fulfil the requirements for B7.2 expression, T cell help and rescue (Cooke et al 1994, and Fig. 2). These would be the circumstances under which the ability to initiate an immune response against a foreign antigen would be traded against low level autoimmunity (Fig. 2). The danger of hypermutation generating higher-affinity self-reacting antibodies is countered by the presence of self-antigen which can abort the survival of high affinity self-reactive cells within germinal centres (Pulendran et al 1995, Shokat & Goodnow 1995).

The biochemical basis of BCR signalling thresholds

The BCR complex is modulated not only by the form and concentration of antigen, but also by the presence of co-receptors such as CD21, CD22 and FcγRIIB1 which are expressed at higher levels as the cells mature. Pathogenic antigens which have

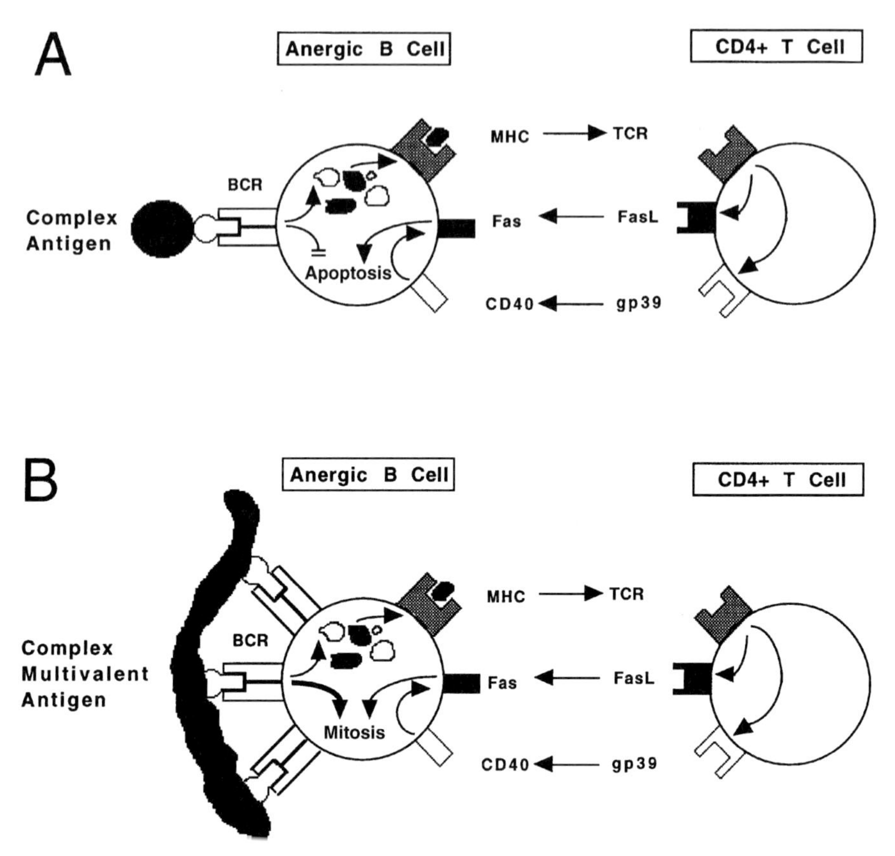

FIG. 2. Anergic B cells can be recruited into an immune response by multivalent foreign antigens. (A) Anergic B cells may interact with competent T cell help by binding complex antigens from which they can present foreign peptide epitopes. Under these circumstances tolerance is normally preserved because the interaction with the T cell leads to Fas-mediated B cell apoptosis. (B) Under other circumstances the foreign antigen may be sufficiently multivalent to overcome the signalling block in anergic cells, and cause an interaction with the T cell which results in mitosis rather than apoptosis. Tolerance to the self antigen is broken in order to generate an immune response against the foreign protein.

been decorated with complement C3d co-cluster the C3d receptor, CD21, with the BCR and enhance calcium signalling and B cell activation through the phosphorylation of the CD21-associated cell surface molecule, CD19, and SH2-mediated recruitment of PI3 kinase and Vav (Dempsey et al 1996). In contrast, B cell activation and BCR calcium signalling is diminished later in an immune response by co-ligation of FcγRIIB1 to the BCR when antigen is decorated by secreted IgG (Daeron 1997). Tyrosine phosphorylation of FcγRIIB1 recruits an inositol polyphosphate 5′-phosphatase, SHIP, to the BCR where it inhibits signal propagation (Ono et al 1997).

Another co-receptor on mature B cells, CD22, binds α2–6 sialic acid on glycoproteins such as IgM and is phosphorylated at both ITAM and ITIM motifs in its cytoplasmic tail after BCR clustering, independent of complement or immune complexes (Cyster & Goodnow 1997). Phosphorylated CD22 binds Syk, phospholipase C-γ, and the phosphatase SHP-1 (Campbell & Klinman 1995, Doody et al 1995, Law et al 1996). Targeted deletion of the CD22 gene has shown that it has a primarily negative role in BCR signalling (Nitschke et al 1997, O'Keefe et al 1996, Otipoby et al 1996, Sato et al 1996), presumably through the recruitment of SHP-1. However, it may also have a positive role, as judged from the variable effect of its deletion on antigen-induced B cell proliferation, and the increase in signalling which occurs when it is independently co-ligated. Interestingly, in the absence of antigen, 0.2–2% of surface IgM is associated with CD22. Therefore the function of CD22 may be to regulate the constitutive level of BCR signalling. It may also integrate undefined environmental cues through binding to its external receptor.

As well as environmental cues, the antigen receptor adapts to persistent signalling by receptor modulation. Low concentrations of sHEL that are below the threshold for inducing anergy nevertheless cause a 50–80% decrease in surface IgM on mature cells which is antigen-dose dependent (Akkaraju et al 1997, Goodnow et al 1989). Increased signalling due to loss of the SHP-1 phosphatase causes spontaneous IgM modulation in proportion to the loss of functional SHP-1 alleles. Partial or complete deficiency of CD45 diminishes BCR signalling and IgM modulation to differing degrees (Cyster et al 1996); increased modulation also occurs in response to increased signalling from overexpression of CD19 (Sato et al 1997) or loss of CD22. IgM modulation is presumably an adaptive response which buffers continuous increases in signalling. Buffering may be effective because cells which carry heterozygote mutations at CD45 or SHP-1 loci are phenotypically normal despite IgM modulation.

Genetic implications of antigen receptor signalling thresholds

One of the significant challenges facing current research is the identification of genetic variants which encode susceptibility to autoimmune disease. Since antigen receptor signalling is central to the balance of tolerance and immunity, it would not be surprising to find such inherited immune modifiers amongst the BCR-associated proteins, particularly those which have a regulatory role. Mutations which arise within BCR signalling pathways would have a powerful means of their own selection, through changes in an individual's ability to fight infections. Genetic research over the last 10 years has shown that susceptibility to complex autoimmune diseases, such as insulin-dependent diabetes mellitus and lupus erythematosus, is determined by the combined effects of several alleles at several genetic loci (Vyse & Todd 1996). Evidence suggests that variant alleles must be relatively common in human populations, and the continuous variation in genetic risk from individual to individual indicates that these alleles must operate in pathways which encode quantitative traits.

The BCR integrates quantitatively variable signals from antigens and co-receptors and translates them into qualitatively different outcomes which are relevant to autoimmune susceptibility. Genetic experiments with signalling mutants confirm that the inherited thresholds for these outcomes are finely tuned and can be modified by genetic variants. Naturally occurring immune variants are most likely to arise in those components of the BCR which are not only regulatory but also rate limiting because these can be subject to natural selection as dominant effects. The fact that deficiency of single alleles of both CD45 and SHP-1 has an observable effect on signalling by receptor modulation suggests that such limiting components may not be infrequent. The strategy of defining the limiting components of BCR signalling will reveal potential sites for inherited immune variants as well as possible targets for therapeutic intervention.

Acknowledgement

R. J. Cornall is supported by a Wellcome Trust Career Development Award.

References

Adelstein S, Pritchard-Briscoe H, Anderson TA et al 1991 Induction of self-tolerance in T cells but not B cells of transgenic mice expressing little self antigen. Science 251:1223–1225

Akkaraju S, Canaan K, Goodnow CC 1997 Self-reactive B cells are not eliminated or inactivated by autoantigen expressed on thyroid epithelial cells. J Exp Med 186:2005–2012

Bell SE, Goodnow CC 1994 A selective defect in IgM antigen receptor synthesis and transport causes loss of cell surface IgM expression on tolerant B lymphocytes. EMBO J 13:816–826

Campbell MA, Klinman NR 1995 Phosphotyrosine-dependent association between CD22 and the protein tyrosine phosphatase 1C. Eur J Immunol 25:1573–1579

Chan EY-T, MacLennan ICM 1993 Only a small proportion of splenic B cells in adults are short-lived virgin cells. Eur J Immunol 23:357–363

Chen C, Nagy Z, Radic MZ et al 1995 The site and stage of anti-DNA B-cell deletion. Nature 373:252–255

Cooke MP, Heath AW, Shokat KM et al 1994 Immunoglobulin signal transduction guides the specificity of B cell–T cell interactions and is blocked in tolerant self-reactive B cells. J Exp Med 179:425–438

Cyster JG, Goodnow CC 1995a Protein tyrosine phosphatase 1C negatively regulates antigen receptor signalling in B lymphocytes and determines thresholds for negative selection. Immunity 2:13–24

Cyster JG, Goodnow CC 1995b Antigen-induced exclusion from follicles and anergy are separate and complementary processes that influence peripheral B cell fate. Immunity 3:691–701

Cyster JG, Goodnow CC 1997 Tuning antigen receptor signalling by CD22: integrating cues from antigens and the microenvironment. Immunity 6:509–517

Cyster JG, Hartley SB, Goodnow CC 1994 Competition for follicular niches excludes self-reactive cells from the recirculating B-cell repertoire. Nature 371:389–395

Cyster JG, Healy JI, Kishihara K, Mak TW, Thomas ML, Goodnow CC 1996 CD45 sets thresholds for negative and positive selection of B lymphocytes. Nature 381:325–328

Daeron M 1997 Fc receptor biology. Annu Rev Immunol 15:203–234

DeFranco AL 1987 Molecular aspects of B-lymphocyte activation. Annu Rev Cell Biol 3:143–178

Dempsey PW, Allison MED, Akkaraju S, Goodnow CC, Fearon DT 1996 C3d of complement as a molecular adjuvant: bridging innate and acquired immunity. Science 271:348–350

Doody GM, Justement LB, Delibrias CC et al 1995 A role in B cell activation for CD22 and the protein tyrosine phosphatase SHP. Science 269:242–244

Goodnow CC 1996 Balancing immunity and tolerance: deleting and tuning lymphocyte repertoires. Proc Natl Acad Sci USA 93:2264–2271

Goodnow CC, Crosbie J, Adelstein S et al 1988 Altered immunoglobulin expression and functional silencing of self-reactive B lymphocytes in transgenic mice. Nature 334:676–682

Goodnow CC, Crosbie J, Jorgensen H, Brink RA, Basten A 1989 Induction of self-tolerance in mature peripheral B lymphocytes. Nature 342:385–391

Goodnow C, Cyster J, Hartley S et al 1995 Self-tolerance checkpoints in B lymphocyte development. Adv Immunol 59:279–368

Hartley SB, Crosbie J, Brink R, Kantor AA, Basten A, Goodnow CC 1991 Elimination from peripheral lymphoid tissues of self-reactive B lymphocytes recognising membrane-bound antigens. Nature 353:765–768

Hartley SB, Cooke MP, Fulcher DA et al 1993 Elimination of self-reactive B lymphocytes proceeds in two stages: arrested development and cell death. Cell 72:325–335

Healy JI, Dolmetsch RE, Timmerman LA et al 1997 Different nuclear signals are activated by the B cell receptor during positive versus negative signalling. Immunity 6:419–428

Kurosaki T 1997 Molecular mechanisms in B cell antigen receptor signalling. Curr Opin Immunol 9:309–318

Law C-L, Sidorenko SP, Chandran KA et al 1996 CD22 associates with protein tyrosine phosphatase 1C, Syk and phospholipase C-γ1 upon B cell activation. J Exp Med 183:547–560

Lortan JE, Roobottom CA, Oldfield S, MacLennan ICM 1987 Newly produced virgin B cells migrate to secondary lymphoid organs but their capacity to enter follicles is restricted. Eur J Immunol 17:1311–1316

Neel BG 1997 Role of phosphatases in lymphocyte activation. Curr Opin Immunol 9:405–420

Nemazee DA, Bürki K 1989 Clonal deletion of B lymphocytes in a transgenic mouse bearing anti-MHC class I antibody genes. Nature 337:562–566

Nitschke L, Carsetti R, Ocker B, Kohler G, Lamers MC 1997 CD22 is a negative regulator of B cell receptor signalling. Curr Biol 7:133–143

O'Keefe TL, Williams GT, Davies SL, Neuberger MS 1996 Hyperresponsive B cells in CD22-deficient mice. Science 274:798–801

Ono M, Okada H, Bolland S, Yanagi S, Kurosaki T, Ravetch JV 1997 Deletion of SHIP or SHP-1 reveals two distinct pathways for inhibitory signalling. Cell 90:293–301

Otipoby KL, Andersson KB, Draves KE et al 1996 CD22 regulates thymus independent responses and the lifespan of B cells. Nature 384:634–637

Pulendran B, Kannaroukis G, Nouri S, Smith KG, Nossal GJV 1995 Soluble antigen can cause apoptosis of germinal center B cells: a model for clonal deletion within the germinal center. Nature 375:331–334

Rathmell JC, Cooke MP, Ho WY et al 1995 CD95 (Fas)-dependent elimination of self-reactive B cells upon interaction with CD4$^+$ T cells. Nature 376:181–184

Sato S, Miller AS, Inaoki M et al 1996 CD22 is both a positive and a negative regulator of B lymphocyte antigen receptor signal transduction: altered signalling in CD22-deficient mice. Immunity 5:551–562

Sato S, Steeber DA, Jansen PJ, Tedder TF 1997 CD19 expression levels regulate B lymphocyte development: human CD19 restores normal function in mice lacking endogenous CD19. J Immunol 158:4662–4669

Shokat KM, Goodnow CC 1995 Antigen-induced B cell death and elimination during germinal-centre immune responses. Nature 375:334–338

Vyse TJ, Todd JA 1996 Genetic analysis of autoimmune disease. Cell 85:311–318

DISCUSSION

Stockinger: Chris Goodnow used to argue that anergic B cells are on the way to die: they don't hang around long. I found this quite reassuring. Are you contradicting this?

Cornall: When anergic B cells develop in the absence of naïve B cells, they recirculate through the secondary lymphoid organs for about two weeks. Therefore the intrinsic lifespan of anergic cells is quite long, albeit less than that of naïve cells. However, in the presence of a mixed repertoire of cells, anergic B cells are unable to gain access to the B cell follicles and die in the T cell zones within about three days. It is interesting to speculate upon why the self-reactive cells should be allowed into the periphery even for a short period. We believe this is most likely a feature of the underlying balance between immunity and tolerance. This is because under certain circumstances when the anergic cells interact with a particularly multivalent or complement-decorated antigen, the advantage of recruiting the self-reactive cells into an immune response would outweigh the danger of autoimmunity. Of course there may be other roles for anergic cells, such as the induction of tolerance in other cells, but the importance of these is also unclear.

Lechler: Since your tolerance phenomena are so beautifully quantitative and you are right at a threshold between activation and tolerance, do you see parallels of T cell observations in which non-responsiveness is achieved by a reduction in the levels of CD4, CD8 or T cell receptor/CD_3 expression? The parallel would be reduced expression of surface Ig or CD40 expression on the B cells of the tolerant animals. Because if you observe a fivefold difference in antigen concentration being critical in your experiments, it wouldn't be very difficult for a B cell to escape deletion simply by regulating either surface Ig or some other co-receptor molecule. Do you see that?

Cornall: I agree. Quantitative differences in signalling and buffering of the sort you suggest might easily allow cells to escape deletion or other tolerance checkpoints. Presumably qualitative differences in cell fate are determined by discrete thresholds on a background of quantitative adjustment. This is also what you'd expect from the genetics of autoimmune disease in which qualitative phenotypes are determined by quantitative trait loci.

Lechler: I'm talking about cells coming through that are not anergic—they're actually potentially responsive—it's just that their threshold for responding is higher because they've down-regulated a critical receptor.

Cornall: It's possible.

Mitchison: I wonder if Robert Lechler, or perhaps Thomas Kamradt, in speaking from their more clinical environments, would have anything to say about defects in B cell tolerance which might lead to disease. Is lupus for example, regarded as a candidate B cell tolerance interference?

Kamradt: Lupus is characterized by autoantibody production, most prominently against nuclear constituents. However, it is clear that the autoantibody production is T cell dependent and there is no good evidence for defects in central B cell tolerance being important for the pathogenesis of systemic lupus erythematosus. Given the association of lupus with certain MHC molecules and the need for T cell help for autoantibody production, I think that T cell tolerance would be a much more promising approach to immunomodulation in lupus than B cell tolerance.

Abbas: Many of us have emphasized the role of the Fas pathway in maintaining T cell tolerance. However, if normal Fas is expressed in the B cells of *lpr* mice, the autoimmune disease is cured, but not the lymphoproliferation. Conversely, putting Fas into the T cell compartment does not cure the autoimmune disease. This is therefore an example where the absence of Fas in B cells is absolutely critical for autoantibody production, presumably because anergic B cells are normally eliminated by the Fas pathway, and they're not getting eliminated and are somehow getting triggered so that they start making autoantibody.

Mitchison: That fits reasonably well with the evidence for Fas-resistant B cells (CD5 B cells) in NZB hybrid mice (Hirose et al 1997).

Cornall: One of the current problems in the analysis of complex diseases is that the unknown susceptibility genes are still linked to very large genomic regions. Ward Wakeland's group in Florida have approached this problem for three murine NZB/NZW lupus susceptibility genes by creating recombinant inbred strains each containing single susceptibility loci (Morel et al 1997, Mohan et al 1997). Using this approach he has been able to associate phenotypes, such as anti-nuclear antibody production, hypergammaglobulinaemia and glomerulonephritis, with separate loci. Interestingly, one of these loci is associated with B cell hyperactivity, which is very similar to that seen in motheaten mice.

Abbas: You showed a map from John Todd. Two of those *Idd* loci turn out to be within 0.5 centimorgans of the IL-2 gene. One is next to the IL-2 gene and the other is within 1 centimorgan of CTLA-4.

Mitchison: That fits with the Colucci et al (1997) data in mice for *Idd5*.

References

Colucci F, Bergman ML, Penha-Goncalves C, Cilio CM, Holmberg D 1997 Apoptosis resistance of nonobese diabetic peripheral lymphocytes linked to the Idd5 diabetes susceptibility region. Proc Natl Acad Sci USA 94:8670–8674

Hirose S, Yan K, Abe M et al 1997 Precursor B cells for autoantibody production in genomically Fas-intact autoimmune disease are not subject to Fas-mediated immune elimination. Proc Natl Acad Sci USA 94:9291–9295

Mohan C, Morel L, Yang P, Wakeland EK 1997 Genetic dissection of systemic lupus erythematosus pathogenesis: Sle2 on murine chromosome 4 leads to B cell hyperactivity. J Immunol 159:454–465

Morel L, Mohan C, Yu Y et al 1997 Functional dissection of systemic lupus erythematosus using congenic mouse strains. J Immunol 158:6019–6028

General discussion I

Abbas: Does B cell tolerance to self proteins occur normally, and is it important? Are our B cells tolerant to our own protein antigens, or is tolerance to protein antigens entirely maintained at the T cell level?

Healy: This is an important question that is difficult to address. *Ex vivo* lipopolysaccharide (LPS) stimulation of B cells from healthy individuals induces plasma cell differentiation and detectable autoantibody production. This demonstrates that autoreactive B cell clones exist in an interactive state at some low frequency. T cell-deficient people such as AIDS patients do not uniformly develop frank autoimmune disease. Although there are rheumatic manifestations in AIDS they do not typically develop B cell-dependent autoimmunity. This would suggest that B cell tolerance can be maintained even with vastly reduced numbers of $CD4^+$ T cells.

Hafler: There is an increase of things such as myasthenia gravis and Guillain–Barré syndrome early on in AIDS patients when there is some CD4 dysregulation.

Healy: Although that may be true, there are many autoantibody-associated autoimmune diseases that these patients could potentially develop, but they do not.

Lenardo: But they're immunodeficient: if you can't generate an immune response, you can't generate autoimmune disease.

Healy: They are T cell deficient. In intravenous drug users who have AIDS, frequent bacterial sepsis produces endotoxins such as LPS, a polyclonal B cell activator that can stimulate rheumatoid factor, anti-DNA antibodies and anti-erythrocyte antibodies. Despite this, B cell-mediated autoimmune diseases do not typically appear, suggesting that B cells have an intrinsic ability for self–non-self recognition.

Hafler: It's such a complex system it's hard to make those extrapolations. There is an increased frequency of autoimmunity in the early stages in patients with HIV. I don't know exactly the data, but certainly myasthenia gravis and Guillain–Barré syndrome occur early on in the disease.

Healy: Myasthenia gravis is an HLA DR3-associated disease that must involve T cell dysregulation.

Stockinger: But surely there's no example of an autoimmune B cell which isn't paralleled by an autoimmune T cell.

Healy: In our lysozyme transgenic system, T cells are not required to induce or maintain B cell tolerance.

Hafler: What about the myelin-associated glycoprotein (MAG) neuropathy? It is a severe autoimmune disease mediated by IgM autoantibodies against the myelin protein on nerves, causing a severe demyelinating neuropathy. This may not be related to T cells at all.

Mitchison: That has an interesting rat model doesn't it?

Hafler: No. That is MOG (myelin oligodendrocyte protein), which induces a different disease. MOG is the CNS disease in which there is a breakdown of the blood–brain barrier and anti-MOG antibodies cause frank demyelination in the CNS. The MAG neuropathy is a naturally occurring disease in humans seen in older males, with slow progressive neuropathy with high titres (one in a million) of this autoantibody.

Mitchison: But the MOG neuropathy, at least in the animal model, is fairly clearly Th2 driven.

Hafler: The recent Hauser paper suggests that MOG-induced experimental autoimmune encephalitis (EAE) may be Th2 driven, but that's not entirely clear (Genain et al 1996). MAG neuropathy is completely different: it's a naturally occurring human disease.

Mitchison: Nevertheless, the MOG model is also thought to refer to a subgroup of multiple sclerosis (MS) patients.

Hafler: No one has been able to demonstrate anti-MOG antibodies in humans with MS. There are MOG-reactive T cells, but no anti-MOG antibodies.

Mitchison: That may be, but as I recall the work of the Wekerle group, they draw a parallel between the MOG-induced disease and an MS subset which may well be directed against a different protein.

Hafler: They do. This is work by Chris Linington, and he has shown very clearly that anti-MOG antibodies which are T cell dependent, and probably also a Th2 response, are involved in demyelination. The pathological picture with the demyelination in the animal models looks like MS. However, MS patients don't have anti-MOG antibodies, suggesting that there may be another antigen, as yet unidentified, which antibodies are directed to leading to demyelination.

Wraith: How clear is it in humans that progression of IgM-mediated diseases is T cell independent? Is MAG neuropathy really a T-independent disease?

Hafler: It is a very important question, but I don't know the answer.

Shevach: One classic model for a T cell-dependent antibody-mediated autoimmune disease is the NZB/NZW F1 mouse. These mice are cured by anti-CD4 treatment (Wofsy & Seaman 1985).

Waldmann: Abul, you questioned whether there is evidence of tolerance for protein antigens in the B cell compartment independent of T cells. There is a classic example which Nossal (1983) quoted in his review, which is to take B cells before they express receptor, polyclonally activate them and show that they can make anti-cytochrome antibodies. This shows that they're competent to do that.

You then take another population that have matured to the stage where they have receptor, polyclonally activate them, and now they don't make anti-cytochrome. He extrapolates that tolerance must have occurred to cytochromes. This seems a very good example: it is a natural situation, unlike putting class I on the surface of cells.

Stockinger: But there the T cells would also be tolerant to cytochrome. I thought we were looking for an example where only the B cells get tolerized and the T cells are unaffected.

Mitchison: The question is whether there's evidence of natural tolerance in the B cell compartment, irrespective of what happens with the T cells.

Abbas: The reason why this question of selective tolerance in either cell population is important is that for a long time people used to say that the NZB systemic lupus erythematosus (SLE) models were due to B cell hyper-reactivity or failure of B cell tolerance. More recently, the evidence for self nucleoprotein-specific helper T cells has increased, and even in the NZB/NZW models people are increasingly saying that it is a primary Th cell defect. Then the question becomes: do you ever have B cell tolerance, or could all tolerance against protein antigens be at the level of T cell help? Obviously, this is a therapeutically important question. Nobody has gone to the trouble of breeding-in either immunoglobulin or TCR transgenes into NZB/NZW F1s, because it's going to be a complicated breeding step.

Hafler: Is there a fundamental difference between IgM autoantibodies and IgG autoantibodies? Many of the human diseases which are truly autoantibody mediated, except for lupus, involve IgM.

Waldmann: The trouble is that to get the pathogenic effect there has to be T cell help to get high affinity antibodies.

Cornall: An elegant model of B cell autoreactivity which does not seem to require T cell help is that descried by Honjo and colleagues (Murakami et al 1992). They made a transgenic mouse expressing a self-reactive anti-erythrocyte antibody derived from NZB mice. This transgene predisposes towards the formation of peritoneal B1 cells, which are hidden from autoantigen and neither eliminated nor anergized unless self-erythrocytes are introduced artificially into the peritoneal cavity. Although there is likely to be T cell tolerance to the erythrocyte antigen, half the mice develop spontaneous haemolytic anaemia associated with the activation of the B1 cells. Honjo has also shown that autoimmune disease can be triggered by the oral administration of LPS (Murakami et al 1994). Therefore enteric T cell-independent antigens or polyclonal activators might circumvent T cell tolerance and cause autoimmune disease in humans.

Mitchison: That is a good point. If you take the Fas-intact autoimmune mice— of which I guess the NZB mice and their hybrids are the best example and where their lupus is at least a reasonable candidate for B cell-driven autoimmunity — if you could isolate the genetic component which would take with it the known B cell defect (for example, the resistance of CD5 B cells to apoptosis) and this generated the autoimmunity, you would be in a very strong position to answer that question. The fact is, the experiment hasn't been done yet.

Cornall: Ward Wakeland's group in Florida are attempting to look at this sort of thing by creating congenic strains of NZW and NZB which carry separate susceptibility loci. They then intend to analyse the effects of separate loci in isolation and hope to map the genes.

Mitchison: I would like to take up a point which Abul Abbas touched on in his presentation this morning, about the total lymphocyte compartment — the 'lymphon' as some people call it. This expands in the Bcl-2 transgenics. Do we have any more information about the control of the size of the lymphon? It is a problem which is not in principle different from what controls the size of the kidneys or liver, for instance, and the fact that we immunologists haven't made much progress with our problem is no reflection on us because the nephrologists are still stuck over the kidney. But we need to say something sensible about it sooner or later.

Abbas: All we know is that in the mice that overexpress Bcl-2 on T cells, the thymus is about twice normal size. In the TCR transgenics (remember, this is a class II-restricted TCR) there is a disproportionate increase in the number of CD8 cells, which are virtually non-existent in the wild-type TCR transgenic. It is as if cells that are not being positively selected are being persuaded to live longer. The spleens and lymph nodes are about twice normal size. An interesting question is why they aren't 10-fold normal size. This is one control on the size of the lymphoid pool, but I don't think anybody really knows what the other ones are.

Lenardo: If you cross a Bcl-2 mouse to an *lpr*, the spleens and lymph nodes are 10-fold larger. This has been done by two groups (Reap et al 1995, Strasser et al 1995). This illustrates two independent influences of controlling the size of the lymphoid pool.

Abbas: That is true, and it exaggerates the disease, which is much worse in these mice. On the other hand, the Bcl-2 over-expressors themselves, at least in some backgrounds, do make autoantibodies, but as is the case with the Lyn-deficient mice Richard Cornall talked about, it's not because tolerance fails, it's just because all antibody-producing cells seem to live a little bit longer — at least, this is what Andreas Strasser believes. It is not really autoimmunity, it's just lots more antibody being made including autoantibodies. But you're right, if you cross the Bcl-2 mice with *lpr* mice, the lymph nodes are huge.

Mitchison: In a sense that is not a very interesting finding, because it just shows that the lymphon is no different from any other organ. The final control on size is likely to be through cell death rather than through cell proliferation. That is of course an enormously important generalization and a fairly new one. Ten years ago people thought that the control of organ size is a control of growth, not of cell death, but now opinion has turned right round. The size of the liver, for example, is believed to be controlled pretty much exclusively by cell death.

Hafler: Certainly, the developing nervous system is controlled by cell death.

Jenkins: Still, there are stromal cell elements controlling that as well. In a lot of the current knockouts of various inflammatory cytokines there are no lymph nodes or white pulp. We need to understand how lymphocytes interact with the non-lymphoid stromal components of lymphoid tissues.

Kioussis: How relevant is your question to that often-used term 'space'? There used to be a notion that when cells are transferred they expand to occupy the available 'space', but this has recently been questioned. Is that related to your question about what controls the size of the lymphon?

Mitchison: Isn't it rather obvious that if you deplete the immune system via radiation or whatever, and then you transfer a few cells, that they will expand and re-populate?

Kioussis: If you do this in a polyclonal situation they will, but not if it is done in a monoclonal situation using transgenic cells.

Stockinger: This repopulation is entirely antigen driven. If you put a monoclonal T cell population into an empty space where there's no antigen, it will not expand.

Kurts: We have obtained completely different results. When we transfer naïve transgenic CD8 T cells, which are specific for ovalbumin, into Rag knockout mice lacking ovalbumin, they proliferate.

Stockinger: Are the CD8 transgenics Rag knockouts?

Kurts: Yes.

Abbas: How long do these cells survive in the absence of antigen?

Kurts: At least 30 days. The speed of expansion is clearly slower in non-transgenic Rag recipients compared with transgenic Rag mice that express ovalbumin. With antigen they expand at a pace of about 5 h per cell cycle; without antigen it takes approximately 1 day per cell cycle.

Hafler: How do you know these mice don't have antigen?

Kurts: We assume the recipient mice do not express any antigen that can be recognized by our transgenic CD8 T cells. This is based on the fact that these cells do not proliferate or show any signs of activation when transferred into non-transgenic C57BL/6 recipients, which have a full repertoire of T cells.

Stockinger: It seems to me either you would have to say there's an intrinsic difference between CD4 and CD8, or if in one system you clearly see there's no expansion, you could also argue there's no cross-reactive antigen whereas in the other system there is. That's the dilemma.

Abbas: Have you put it into a class I knockout mouse?

Kurts: No.

Mitchison: Brigitta, if expansion of a population needs to be driven by antigen, it doesn't mean that the space isn't important, because if you transfer the same cells into a normal mouse they don't expand.

Stockinger: In the normal mouse you have follicular competition and things like that going on.

Mitchison: Isn't that a form of space?

Stockinger: No, it's a form of the status of responsiveness of the cell.

Waldmann: Av Mitchison, in your introduction you highlighted the Zinkernagel experiment as indicating something very important. I wasn't entirely clear what you were driving at. You used the word 'neglect' to cover what could have been one of many possibilities: it could have been that the T cells ignored the antigen, or that the T cells were rapidly regulated so at that level they could not react. What was it that you were trying to highlight?

Mitchison: I was trying to highlight that had this meeting taken place 30 years ago, we would have thought about tolerance and immunological disease exclusively in terms of whether there were reactive T cells or not (Simonsen's 'immunologically competent cell'). It is now very clear that there aren't. I was making the additional point that it is still puzzling how as many as 2% of chicken T cells reacted against the chorioallantois when transplanted into an allogenic egg (Nisbet et al 1969).

Waldmann: It must be 20 years ago that Russell transplanted a mouse kidney that differed across an MHC barrier into another animal and there was no rejection. That clearly was setting something in motion, a process leading to tolerance where in other situations, other grafts (e.g. skin) would promote rapid rejection. I look at the Zinkernagel experiment in the same light. It doesn't surprise me that there are circumstances where you can put an antigen into a tissue and you don't get a reaction. The burning question concerns what the basis of this is: is it simply ignorance—as he would like to say—or is there more to it? If you went back and investigated the Russell experiment, I bet you would find regulation.

Mitchison: I think it has much more to do with inflammation and IL-1.

Mason: I suspect it's the nature of the antigen-presenting cell (APC) in the two cases. The APC in the chorioallantois experiment I should imagine is what we would call a dendritic cell today.

Mitchison: I guess most people think it depends on whether APCs are in the mood to trigger a response or not.

Cornall: Just before we finish discussing space, autoimmune disease appears to develop in states where we have immunodeficiency. It's an important point that autoantibodies are often produced in cases where there is immunodeficiency. This may be due to autoreactive cells filling up space that has become available to them under circumstances where they would normally be competed out.

Shevach: There I would fully agree with Herman Waldmann. This is all to do with regulatory T cells: that's what's missing.

Mason: I agree.

Abbas: In fact there's no evidence for that. In the 25% or so of agammaglobulinaemics that develop rheumatoid arthritis there's no evidence that this is because of abnormal regulatory T cells.

Shevach: In the rat and mouse there is evidence.

Abbas: Is there evidence that autoimmunity associated with immunodeficiency states is due to defective regulation?

Mason: The BB rat is an example of that.

Mitchison: I dare say you are right in that there is no conclusive evidence that this represents a disturbance in immunoregulation, but that is what it smells like.

Abbas: To be more specific, what you would say is that it's either a failure of Th2 development or a lack of anti-inflammatory T cells. That's really what you mean by immune regulation.

Shevach: No, that is just one kind of immune regulation; there may be others.

Abbas: So that's the question: is there any evidence that those immunodeficiencies are associated with preferential or selective defects in suppressor cells or anti-inflammatory T cells?

Shevach: That's very hard to study in humans: we don't have good markers for these cells in rodent models yet, so in the human situation that's impossible. But there are T cell-mediated diseases that also develop. For example, 10% of patients with common variable immunodeficiency develop pernicious anaemia, a T cell-mediated autoreactive disease destroying the gastric mucosa (Eisenstein & Sneller 1994).

Mitchison: Don Mason, you must have an overview about that, because you've been studying T cell markers as predictors of autoimmunity. I guess it's true that most of the immune defects are not present in rats so you may have to go and look in mice.

Mason: The BB rat is genetically immunodeficient. It has a paucity of T cells and it gets autoimmune diabetes. There are congeneic lines which are not lymphopenic, and you can transfer cells from those animals into the lymphopenic line and prevent disease. This is therefore a genetic defect which leads to lymphopenia which leads to autoimmunity and is curable if you put in the right population of cells.

Waldmann: Can I give you one clinical example. It involves treatment of MS patients with an anti-lymphocyte antibody that depletes T cells. Certainly, there's evidence that acute lesions of MS are benefited, but about a third of these patients get thyroid autoimmune features: either frank thyrotoxicosis or anti-thyroid antibodies. This is a surprising outcome of lymphocyte depletion.

Cornall: Some of the suppressor phenomena may be determined more by the nature of the suppressed cell than by the suppressor itself. For example, we know that anergic B cells are eliminated by immunocompetent T cells with which they interact by a Fas-dependent mechanism (Rathmell et al 1995). The same T cells can help naïve cells which bind to microbial antigen, because these B cells are able to signal from their antigen receptors.

References

Eisenstein EM, Sneller MC 1994 Common variable immunodeficiency: diagnosis and management. Ann Allergy 73:285–292

Genain CP, Abel K, Belmar N et al 1996 Late complications of immune deviation therapy in a nonhuman primate. Science 274:2054–2057

Murakami M, Tsubata T, Okamoto A et al 1992 Antigen-induced apoptotic death of Ly-1 B cells responsible for autoimmune disease in transgenic mice. Nature 357:77–80

Murakami M, Tsubata T, Shikura R et al 1994 Oral administration of lipopolysaccharides activates B-1 cells in the peritoneal cavity and lamina propria of the gut and induces autoimmune symptoms in an autoantibody transgenic mouse. J Exp Med 180:111–121

Nisbet NW, Simonsen M, Zaleski M 1969 The frequency of antigen-sensitive cells in tissue transplantation. A commentary on clonal selection. J Exp Med 129:459–467

Nossal GVJ 1983 Cellular mechanisms of immunological tolerance. Annu Rev Immunol 1: 33–62

Rathmell JC, Cooke MP, Ho WY et al 1995 D95 (Fas)-dependent elimination of self-reactive B cells upon interaction with CD4+ T cells. Nature 376:181–184

Reap EA, Felix NJ, Wolthusen PA, Kotzin BL, Cohen PL, Eisenberg RA 1995 bcl-2 transgenic Lpr mice show profound enhancement of lymphadenopathy. J Immunol 155:5455–5462

Strasser A, Harris AW, Huang DC, Krammer PH, Cory S 1995 Bcl-2 and Fas/APO-1 regulate distinct pathways to lymphocyte apoptosis. EMBO J 14:6136–6147

Wofsy D, Seaman WE 1985 Successful treatment of autoimmunity in NZB/NZW F1 mice with monoclonal antibody to L3T4. J Exp Med 161:378–391

The study of self-tolerance using murine haemoglobin as a model self antigen

Calvin B. Williams and Paul M. Allen

Center for Immunology and Departments of Pathology and Pediatrics, Washington University School of Medicine, 660 S. Euclid Avenue, St Louis, MO 63110, USA

Abstract. T cell tolerance to self proteins involves both thymic and peripheral mechanisms. We have used allotypic differences in murine haemoglobin (Hb) to study the development of tolerance to this abundantly expressed self-protein. In Hbβ^s/H-2^k mice, the response to Hbβ^d is directed against Hbβ^d(64–76) presented by I-E^k molecules. Using T cell hybridomas and clones specific for this epitope, we have demonstrated that Hb(64–76)/I-E^k complexes are present on antigen-presenting cells in all lymphoid organs including dendritic cells, B cells and macrophages. In the thymus, the presence of these complexes results in negative selection of transgenic T cells with high levels of Hb(64–76)/I-E^k-specific receptor. However, cells with intermediate levels of specific receptor escape negative selection and can be found in the periphery. Under normal circumstances these cells remain tolerant, but can be activated by mechanisms which increase the number of Hb(64–76)/I-E^k complexes.

1998 Immunological tolerance. Wiley, Chichester (Novartis Foundation Symposium 215) p 41–53

The key challenge in the development of an anticipatory immune system is to have as broad a T cell repertoire as possible, in the absence of autoreactivity. The self-tolerance of T cells is achieved through several different tolerance mechanisms, involving both central and peripheral tolerance (Kruisbeek & Amsen 1996, Marrack 1993, von Boehmer 1992, Nossal 1994). Over the past 10 years, our laboratory has developed and utilized the self-protein murine haemoglobin as a model self-antigen. From these studies, we have elucidated how the immune system handles haemoglobin, and how tolerance is achieved.

Haemoglobin the protein

Haemoglobin (Hb) is an abundant protein, composed of two pairs of α and β chains (Hardison 1996). It is a circulating self protein, which does not exist free in the circulation. It is found either inside a red blood cell (RBC), or bound to circulating haptoglobin, with free Hb actually being nephrotoxic (Cerda & Oh 1990). The exact concentration of Hb accessible to the immune system is difficult to determine precisely; however, one can make a reasonable calculation. In the cytoplasm of a RBC, Hb is found at the extraordinary concentration of 340 mg/ml. RBCs have a half-life of 120 days. One can then make the calculation that in mice approximately 1–2 mg of Hb per day is degraded and potentially available to the immune system.

There are two allelic forms of the murine Hb β chain, d (diffuse) and s (single) (Russell & McFarland 1974). The d allele expresses two molecules, a Hbβ^{dmajor} and Hbβ^{dminor}, representing 80% and 20% of the Hb, respectively. In addition, there are embryonic forms of Hb as well as allelic forms of the Hbα chain (Wong et al 1986). The sequences of all these Hb molecules have been determined and despite the conserved function, they have many amino acid sequence differences between them (Gilman 1976, Popp 1973, Popp et al 1982). Pertinent to our work, there are 12 amino acid differences between Hbβ^{dminor} and Hbβ^{s}.

Hb(64–76)-specific T cells

We were initially interested in ascertaining whether self proteins were processed and presented by antigen-presenting cells (APCs) *in vivo*. For use as functional probes, we developed T cells specific for Hbβ^{d} by immunizing CE/J mice (H-2^{k}, Hbβ^{s}), with Hbβ^{d} protein. From the strong T cell response we generated a panel of T cell hybridomas and clones which were specific for Hbβ^{d} (Lorenz & Allen 1988a, Evavold et al 1992). Characterization of these T cells revealed that they were specific for the Hbβ^{dminor}(64–76) determinant bound to the I-E^{k} molecule. A comparison of the Hbβ^{d} and Hbβ^{s} alleles revealed that two of the 12 amino acid differences were located in this region. For simplicity, we refer to the Hbβ^{dminor}(64–76) determinant as Hb(64–76). Thus, we had generated T cells in Hbβ^{s} mice, with which we were able to probe Hbβ^{d} mice for the expression of Hb/I-E^{k} complexes on APCs *ex vivo*.

Detection of Hb(64–76)/I-E^{k} complexes *ex vivo*

Using the T cell hybridomas and clones, we were able to show definitively that self-proteins were constitutively processed and presented *in vivo*, as originally proposed by Mitchison and colleagues (Winchester et al 1984). APCs from Hbβ^{d} mice and from all lymphoid organs, spleen, lymph nodes and thymus were able to stimulate the T cells, without the addition of any exogenous antigen (Lorenz & Allen 1988a). We also showed that both professional and non-professional APCs expressed Hb(64–76)/I-E^{k} complexes (Lorenz & Allen 1988a, Hagerty et al 1991, Hagerty

& Allen 1992). These included macrophages, dendritic cells, B cells and kidney proximal tubule cells. These finding directly demonstrated that all APCs constitutively processed Hb and expressed the Hb(64–76)/I-E^k epitope on their surface (Lorenz & Allen 1988b).

Tolerance to Hb

The essential role of Hb in supporting aerobic metabolism clearly indicates that autoreactivity could have serious consequences. Immunization of Hbβ^d mice, with either intact Hbβ^d or Hb(64–76), failed to elicit any detectable T cell response, thus confirming that Hbβ^d mice are tolerant to Hb. This model system was unidirectional in that while Hbβ^s mice respond to Hbβ^d, immunization of Hbβ^d mice with the corresponding Hbβ^s(64–76) peptide did not result in any detectable response. This lack of response was explained by the sequence of Hbβ^s being identical in this region to the Hbβ^{dmajor} sequence and that this peptide also lacks two key residues involved in binding to I-E^k.

We then wanted to explore how this tolerance was generated. For central tolerance, a self-antigen must be expressed in the thymus, where it can cause the elimination of the self-reactive T cells. We examined the thymic APCs for the expression of Hb(64–76)/I-E^k complexes. Surprisingly, we detected the complexes on all of the APCs in the thymus: cortical epithelial cells, cortical/medullary junction dendritic cells, and medullary macrophage and dendritic cells (Lorenz & Allen 1989a,b). It had been proposed that one way for positive selection to occur would be to have a limited or different set of self peptides expressed in the cortex, compared with the medulla. This observation showed that self proteins were accessible to the cortex and that there was not a true blood/thymus barrier. We contend that the tolerance to Hb is achieved by negative selection in the thymus of the Hb-reactive T cells. Our findings indicated that the complexes were detected both in the cortex and the medulla, raising the possibility that negative selection was occurring in both sites.

Mature T cells begin to develop as early as embryonic day 15 (E15). We examined when we could detect the Hb(64–76)/I-E^k complexes in the fetal thymus. We detected complexes as early as day E14, the level of which increased until birth (Lorenz & Allen 1989b). Thus, the Hb(64–76)/I-E^k complexes exist prior to the development of T cells, and would be available for deletion of the self-reactive T cells.

Hb tolerance using a Hb TCR transgenic system

To examine the tolerance to Hb at a single cell level we have recently generated a TCR transgenic mouse using the TCR chains from the 3.L2 T cell clone (Kersh et al

1998). These mice were produced in the B6 background, and then bred to B6.AKR(H-2^k) mice to introduce the I-E^k restriction element, and to maintain homozygosity of the Hb^s allele. The cells expressing the 3.L2 TCR can be detected by a clonotypic antibody (CAB). T cells expressing the 3.L2 TCR were selected into the CD4 lineage in the thymus, comprised 50% of the $CD4^+$ T cells in the periphery, and were strongly reactive to Hb(64–76).

To observe the process of tolerance of the 3.L2 T cells, we bred the 3.L2tg mice to a B6.Hb^d congenic line, thus producing mice which only differ by the presence or absence of the self antigen. In the thymus of these mice, the CAB^{hi} $CD4^+$ T cells were eliminated, but the CAB^{int} cells did still exist. Thus, the presence of the self antigen induced deletion of the Hb(64–76) reactive cells with high levels of the 3.L2 TCR. Interestingly, the $CD4^+CD8^+$ cells in the cortex were not deleted, even though Hb(64–76)/I-E^k complexes were present. Similar conclusions were made in the study of the self antigen C5 by Stockinger and colleagues (Lin & Stockinger 1989, Stockinger et al 1993, Zal et al 1994). Our results showed that for a unmanipulated self-antigen, central tolerance was occurring at the $CD4^+$ single-positive stage, at the cortical medullary junction and/or the medulla.

In the periphery, there are a few but detectable number of CAB^+ cells. These T cells did react to Hb(64–76), but at much higher concentrations. Their presence indicated that the tolerance to Hb(64–76) was not absolute. These cells expressed low levels of TCR, making their reactivity sub-threshold for the endogenous level of Hb(64–76)/I-E^k found in the periphery. These are therefore potentially self-reactive T cells, which under the appropriate set of conditions could be stimulated, resulting in autoimmunity. The use of the 3.L2tg mouse has allowed us to visualize the negative selection of the Hb-reactive T cells in the thymus, to demonstrate that tolerance to Hb is not absolute, and that subthreshold self-reactive T cells are part of the peripheral T cell pool.

Effect of an endogenous altered peptide ligand

We have also expressed an altered peptide ligand (APL) of Hb(64–76) in transgenic mice and followed its effect on 3.L2 T cell development in the thymus and peripheral responses (Williams et al 1998). The APL, A72, is a TCR antagonist, and was found to have minimal effects on the development of thymocytes. Surprisingly, it was shown to antagonize peripheral responses. Therefore, the A72 ligand did not negatively select the 3.L2 T cells, but it did antagonize their response. These results indicate that there could be a constitutive level of antagonism by self-peptides, whose effects can be overcome by more antigen. By not deleting these reactive T cells, the T cell repertoire size is maintained, while the antagonism would prevent any potential autoreactivity.

Concluding remarks

The Hb self-antigen system is an ideal model for the study of the process of self-tolerance in a unmanipulated system. Through its use, we now know that self-proteins are constitutively processed and presented by all APCs in the host. The Hb/Ia complexes expressed in the thymus are critical to the process of self-tolerance, resulting in the elimination of the self-reactive Hb T cells. This process does not delete all of the potentially reactive cells, but only those which can productively interact. Thus, there are self-Hb reactive T cells present in the periphery; however, the level of Hb/Ia complexes is insufficient under normal circumstances to stimulate these T cells. The existence of these T cells does, however, make it possible that under the correct combination of events, they could be activated, causing an autoimmune reaction. These findings also highlight the delicate balancing act the immune system has to perform in generating a broad enough repertoire to recognize any potential pathogens, while maintaining a lack of self-reactivity.

References

Cerda S, Oh SK 1990 Methods to quantitate human haptoglobin by complexation with hemoglobin. J Immunol Methods 134:51–59

Evavold BD, Williams SG, Hsu BL, Buus S, Allen PM 1992 Complete dissection of the Hb(64–76) determinant using Th1, Th2 clones and T cell hybridomas. J Immunol 148:347–353

Gilman JG 1976 Mouse hemoglobin beta chains. Comparative sequence data on adult major and minor beta chains from two species, *Mus musculus* and *Mus cervicolor*. Biochem J 159:43–53

Hagerty DT, Allen PM 1992 Processing and presentation of self and foreign antigens by the renal proximal tubule. J Immunol 148:2324–2330

Hagerty DT, Evavold BD, Allen PM 1991 The processing and presentation of the self-antigen hemoglobin: self-reactivity can be limited by antigen availability and co-stimulator expression. J Immunol 147:3282–3288

Hardison RC 1996 A brief history of hemoglobins: plant, animal, protist and bacteria. Proc Natl Acad Sci USA 93:5675–5679

Kersh GJ, Donermeyer DL, Frederick KE, White JW, Hsu BL, Allen PM 1998 TCR transgenic mice in which the usage of transgenic alpha and beta chains is highly dependent on the level of selecting ligand. J Immunol, in press

Kruisbeek AM, Amsen D 1996 Mechanisms underlying T-cell tolerance. Curr Opin Immunol 8:233–244

Lin RH, Stockinger B 1989 T cell immunity or tolerance as a consequence of self antigen presentation. Eur J Immunol 19:105–110

Lorenz RG, Allen PM 1988a Direct evidence for functional self protein/Ia-molecule complexes *in vivo*. Proc Natl Acad Sci USA 85:5220–5223

Lorenz RG, Allen PM 1988b Processing and presentation of self proteins. Immunol Rev 106:115–127

Lorenz RG, Allen PM 1989a Thymic cortical epithelial cells lack full capacity for antigen presentation. Nature 340:557–559

Lorenz RG, Allen PM 1989b Thymic cortical epithelial cells can present self antigens *in vivo*. Nature 337:560–562
Marrack P 1993 T cell tolerance. Harvey Lect 89:147–155
Nossal GJV 1994 Negative selection of lymphocytes. Cell 76:229–239
Popp RA 1973 Sequence of amino acids in the β chain of single hemoglobins from C57BL, SWR and NB mice. Biochim Biophys Acta 303:52–60
Popp RA, Bailiff EG, Skow LC, Whitney JB 1982 The primary structure of genetic variants of mouse hemoglobin. Biochem Genet 20:199–208
Russell ES, McFarland EC 1974 Genetics of mouse hemoglobins. Ann N Y Acad Sci 241:25–38
Stockinger B, Grant CF, Hausmann B 1993 Localization of self antigen: implications for antigen presentation and induction of tolerance. Eur J Immunol 23:6–11
von Boehmer H 1992 Thymic selection: a matter of life and death. Immunol Today 13:454–458
Williams CB, Vidal K, Donermeyer D, Peterson DA, White JM, Allen PM 1998 *In vivo* expression of a T cell receptor agonist: T cells escape central tolerance but are antagonized in the periphery. J Immunol, in press
Winchester G, Sunshine GH, Nardi N, Mitchison NA 1984 Antigen-presenting cells do not discriminate between self and nonself. Immunogenetics 19:487–491
Wong PMC, Chung S-W, Reicheld SM, Chu DHK 1986 Hemoglobin switching during murine embryonic development: evidence for two populations of embryonic erythropoietic progenitor cells. Blood 67:716–721
Zal T, Volkmann A, Stockinger B 1994 Mechanisms of tolerance induction in major histocompatibility complex class II-restricted T cells specific for a blood-borne self-antigen. J Exp Med 180:2089–2099

DISCUSSION

Abbas: You said that you thought the altered peptide ligand (APL) was not having much of an effect on the thymus. In fact, what you're seeing is deletion of all the high TCR-expressing cells in the thymus, and those that are left may not be responsive. In other words, the APL may induce central tolerance.

Allen: When we look at the receptor levels in the periphery, we can see that it has been compensated in some mice. We do not at this time understand how the levels have been increased in the periphery. It could be due to some up-regulation in the periphery or some type of preferential expansion.

Wraith: How do you know whether you are seeing deletion of T cells expressing high-affinity receptors as opposed to down-regulation on those cells?

Allen: We don't. Deletion would be consistent with the data and would fit in with the fact that we're not seeing negative selection. We're trying to examine that now.

Simon: We have obtained completely different results with the expression of a partial agonist in the thymus. We find that the cells don't even get to the double-positive stage: they exit the thymus at the $CD8^{lo}$ immature state, just before the double positives. These cells migrate into the periphery and are functional. This is probably because the transgenic TCR is expressed very early in these transgenic mice.

Allen: Which mice are you using?

Simon: K^b-specific mice: BM3.3, not the Désirè mice.

Allen: Our transgenic TCR is definitely not expressed early. We don't get robust positive selection. This difference may be due to the mouse models we use.

Allison: TCR $\alpha\beta$ genes that are expressed early in development may appear on cells that are in fact $\gamma\delta$ T cells. It is therefore dangerous to extrapolate from data such as these anything about $\alpha\beta$ T cell development, since the $\gamma\delta$ cells are not negatively selectable.

Simon: But what we show is that these immature $CD8^{lo}$ cells are cycling, and they are also selected on a Rag background.

Allison: It's hard to say that they can't be. There are no independent lineage markers.

Mitchison: In this connection you said something very surprising, which was that you expected the thymus to be more sensitive to deletion. It is well established that the threshold of deletion is identical in the thymus and periphery (Mitchison 1993).

Allen: There is definitely evidence supporting my claim. Jameson and Bevan examined the question of why a T cell doesn't react to its positive-selecting ligand (Jameson et al 1994). They found that when a T cell sees a positive-selecting ligand, it decreases its CD8 co-receptor level. When that T cell comes out into the periphery and sees that ligand it doesn't react to it. Charlie Janeway has also published that the thymocytes were more sensitive to the same signal than peripheral T cells (Yagi & Janeway 1990).

Mitchison: No, it has been very well looked at, and they are identical.

Wraith: If you take our system, where we have peptides of different affinity, you can produce deletion of double-positive thymocytes without having any effect on single-positive thymocytes and T cells in the periphery. Thus it is quite clear that apoptosis with double-positive thymocytes is incredibly sensitive to the dose of antigen (Liu et al 1995).

Allen: We have the same data in our system, but then our cells don't normally delete there. The question would be: what about a single-positive cell coming through, or right at that junction? Conceptually the idea was that this was a mechanism by which peripheral T cells could develop in order not to recognize the things which they had to interact with in the thymus. It is an appealingly simple mechanism, that if the environment in the thymus was lower threshold, and then you just raised the threshold, this would solve the issue.

Jenkins: So in the presence of the antagonists, are the clonotype-positive cells naïve in phenotype ($CD45^+$ Rb^{hi} L-selectin-positive)?

Allen: We have only looked at this a couple of times, and we haven't seen any markers of activation. But we're doing experiments now in which we transfer the T cells in and try to get to the issue of memory. We are developing mice with a whole series of APLs. We will take the 3.L2 T cells and transfer them into different environments and look at whether we get any of the markers of activation or whether there's an effect on how long they live.

Abbas: Those cells don't look like they are functionally anergic, because you can trigger them easily with native peptide.

Allen: Yes. We have no evidence of the cells being anergic.

Jenkins: Phenotype conversion can be very sensitive. It would be great to see whether there is any TCR occupancy going on in that system.

Allen: We have to go back and look at many more events.

Mitchison: That's a fairly inspiring experiment from the point of view of future therapy. What would you think would happen if you made a haemopoietic chimera in which some cells were expressing the APL and other APCs were not? Do you think you might still be able to turn off all T cells?

Allen: Hb(64–76) is an academic antigen. Thus we have examined the process through the eyes of one T cell. There are other T cells for which A72 is an agonist. Their existence makes treatment with APLs difficult.

Hafler: That's a profound problem going into clinical trials. We have done a lot of work looking at APLs and myelin basic protein (MBP) active clones, and Paul's point is exactly right: a peptide that is an agonist may turn off and induce Th0 to a Th2 clone that works perfectly well in one individual, whereas in another subject the same peptide may have a totally different biological function.

Mitchison: We're not short of clinical situations with oligoclonal T cells. If you introduced in some way a subpopulation of APCs carrying the APL, would you expect them to be able to turn off everything? Or do the *in vitro* experiments already answer that question, and say that they have both got to be in contact with the same T cell at the same time?

Allen: I would say that there wouldn't be an effect, but it would be hard to demonstrate. And I think that they would have to be on the same APC.

Shevach: When you clean up the T cells and put the normal APCs in, what happens when you put back the antagonist? Does it look exactly the same as the controls — the ones that haven't seen the A72 during development?

Allen: We haven't done it; I think it's a good experiment.

Abbas: In the intact transgenics, can you immunize with the APL containing A72 in adjuvant?

Allen: We haven't done that experiment.

Simon: In our transgenic TCR model, T cells that have seen the partial agonist in the thymus then react to the full agonist in the periphery, but with a partial activation programme. Do the T cells which you find in the periphery have similar characteristics?

Allen: Our 'gold standard' generally is that proliferation indicates full activation. However, we haven't done it. We have another system in which when we change a ligand very subtly we get IL-4 versus IFN-γ. I think it's a good point:

perhaps we should go back and see whether there really is some change and this is not the right assay.

Kurts: You explained the slight changes in affinity resulting in such different responses of the T cells, by the formation of TCR multimers. What about serial engagement of TCRs: would that also explain these differences?

Allen: These are all variations on a common theme. I think the serial triggering idea was that you try to engage a TCR, and once it's engaged, then it can attach to the cytoskeleton. Then you disengage the ligand and bring in another one, so you're forming a contact cap. We don't have any data one way or the other. The point I wanted to make was that at least two kinetic events must be occurring, and that they're interrelated but not directly proportionate. We don't know what the relationship is.

Stockinger: With respect to the double TCR expression, I guess these experiments have not yet been done on the Rag background. It seems that the level of the receptor is quite 'plastic', so in the presence of a strong negatively selecting ligand only those cells that have very low levels of the receptor escape. With your partial agonist, it could be that more cells escape that have relatively higher levels of the specific receptor and that this is then modulated more in the periphery by density of the ligand and so on. It will be important to see whether in the Rag background you get any escape of such partially activated cells.

Allen: We are currently breeding those mice; it is just that the I-A^b problem is slowing us down.

Hämmerling: I would like to ask a very general question about APLs. Your neighbour in St Louis, Emil Unanue, has data showing that quite frequently T cell hybridomas that have been generated by immunization with a synthetic peptide epitope will not react with the same epitope generated by processing of the whole protein (Viner et al 1996). I was very pleased to see that you have put your APL in the context of the whole protein. But how do you know the response to this endogenously processed APL is exactly the same as to the APL you add as a peptide? Does the finding by Viner et al complicate the work on APLs?

Allen: That is a good point. The way we have done it is to make T cells against some of these APLs, immunized with these, and then we can compare the reactivity of the cells when the peptide is in a covalent naturally processed form versus when we add the peptide. In this experiment they are equal as agonists. I-E molecules are much less susceptible to these changes than I-A molecules.

Now what we have to do is to make an antibody that can recognize the peptide/MHC complex. We can then quantitate exactly how many complexes there are. Then we would be in a position to compare directly the number of complexes and their biological activity.

Abbas: Is the phenomenon that you showed — that T cell antigen recognition in the thymus is leading to a loss of high receptor expressing cells either by down

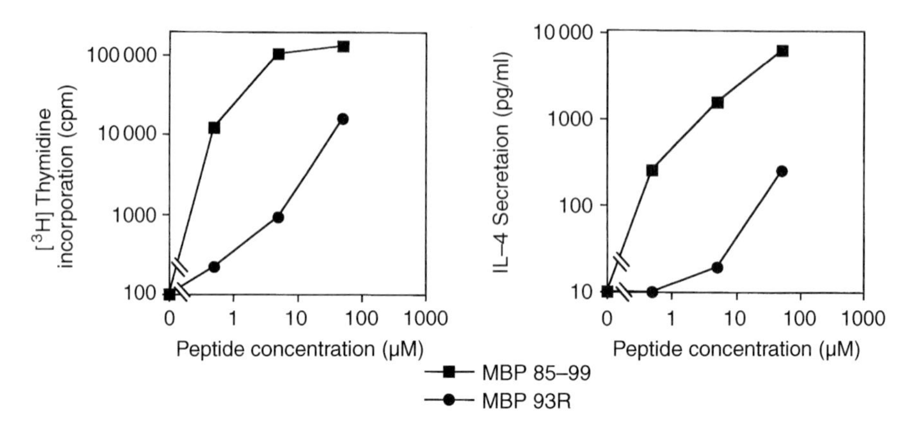

FIG. 1 (*Hafler*) Weak and partial T cell agonists of T cell clones identified by B cell presentation of peptides. An EBV-transformed B lymphoblastoid cell line (MGAR) that is homozygous for DRB1*501 was used to present the indicated peptides to Ob.20b (A), Ob.1A12 (B) and Ho.B27 (C) T cell clones. MBP p85–99 (93R) and MBP p85–99 (93A) are weak agonists of T cell clones Ob.20b and Ob.1A12, respectively, in terms of [^{3}H]TdR incorporation and IL-4 secretion. MBP p85–99 is a partial agonist for Ho.B27, in that it induces IFN-γ secretion in the absence of [^{3}H]TdR incorporation. Data from a representative experiment from each clone are presented, and all experiments were repeated at least three times with similar results.

modulation or deletion that's being compensated for in the periphery — a general one? You clearly see a loss of TCR high cells in the thymus but there is a normal distribution of TCRs in the periphery.

Kioussis: Our experience is that in most antigen/TCR double-transgenic mice the high TCR cells are not found in either the thymus or the periphery; and the few that occur in the periphery have lower levels of TCR.

Abbas: This is breeding which mice?

Kioussis: F5 TCR transgenic mouse with a mouse that expresses nucleoprotein (the cognate antigen).

FIG. 2 (*opposite*) (*Hafler*) Functional differences in B7–1 and B7–2 co-stimulation are seen by presentation of weak, but not agonist, peptides to a human T cell clone. CHO transfectants were fixed for 4 min in 0.2% paraformaldehyde, quenched with 0.1 M L-lysine, washed twice in PBS, and left in culture medium at 37 °C for 16 h. For each sample, transfectants were then incubated for 2 h at 37 °C with the indicated amount of peptide. T cell clone Ho.B27 recognized MBP p85–99 (93R) as its native Ag. Cells were washed, and 5×10^4 CHO cells were incubated with 10^5 T cells/well. Supernatants were collected after 48 h; cytokines were measured by ELISA and T cell proliferation was measured by [^{3}H]TdR incorporation. Cytokines and proliferation were measured in duplicate. Similar results were seen in three independent experiments, and the SEM was <15% for any data point.

IL–5 Secretion (pg/ml)

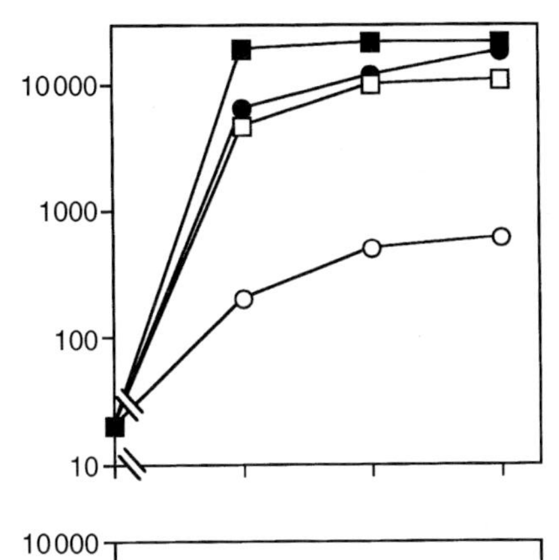

10000
1000
100
10

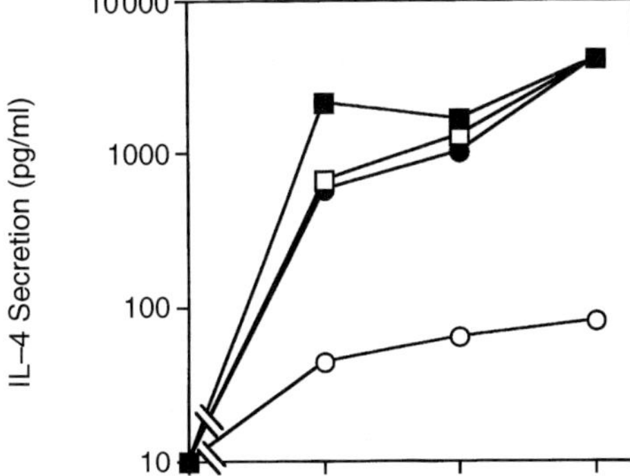

IL–4 Secretion (pg/ml)
10000
1000
100
10

DR2/B7–1 MBP p85–99
DR2/B7–1 MBP p93R
DR2/B7–2 MBP p85–99
DR2/B7–2 MBP p93R

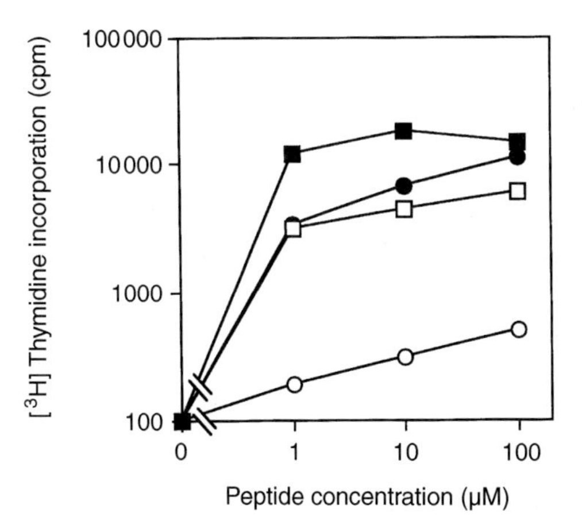

[^{3}H] Thymidine incorporation (cpm)
100000
10000
1000
100
0
1
10
100
Peptide concentration (μM)

Abbas: Have you done that with the mouse that expresses a very low concentration of the nucleoprotein?

Kioussis: In this case we find the opposite effect. In the thymus we have perfect maturation of CD8 cells with high TCR but we don't find them in the periphery. Those few we do find in the periphery have low levels of TCR. Again, this is the opposite of what Paul Allen is finding.

Mitchison: Paul, how sure are you that the thymic mechanism doesn't operate in the periphery as well? How old were the mice you looked at? It would be important to look at old mice (with a relatively small emigration of T cells from the thymus) or, better, at adult thymectomized mice.

Allen: We examined both young and old mice.

Mitchison: Supposing there is some sort of deletion going on mediated by high-expressing cells?

Allen: There could be: that's why I said there could be multiple effects, but the end result is that they look the same.

Mitchison: If they are perhaps cycling in and out of high expression, then you might have to wait for longer than eight weeks to see the effect of a deletional mechanism.

Allen: There could be a very complex phenomenon going on in the thymus. With one of these ligands we think that we may be able to see enhanced positive selection if we get just the right affinity. When you don't see big changes it's hard to be conclusive, especially with the thymus. We've looked at several markers and we don't see any differences in the CD5 or CD69 levels. The scientific community has examined a few TCR transgenics and we're trying to extrapolate that to a whole repertoire: the conclusions we make from these TCR transgenic mouse models will be incomplete. We have to examine many more of these to really figure the whole story.

Kioussis: It is important that Paul Allen has shown that an antagonist can keep the cells under control.

Abbas: But you don't know if they're being kept under control: they're just not responding.

Allen: To do this we have to breed them onto the Rag background.

Hafler: I would like to describe some experiments done by Dave Anderson, a graduate student in my lab (Anderson et al 1997). We have DR2 transfectants expressing either nothing or B7–1 or B7–2, and have used them to present both weak agonists and strong agonists for several human MBP-reactive T cell clones. It is analogous to your experiment using B7–2. Figure 1 (*Hafler*) shows a weak agonist based on a dominant epitope from MBP which has been used to stimulate a T cell clone; I am showing T cell activation using the weak agonist presented by an EBV-transformed B cell line in terms of thymidine incorporation and IL-4 secretion. With a native peptide (MBP 85–99) one gets nice proliferation and IL-4 secretion, and with a weak agonist (MBP 93R) with a substitution in the T cell receptor contact residue there is only a modest response. In Fig. 2 (*Hafler*) we've got the different transfectants expressing B7–1 versus B7–2, with the native peptide giving the full signal when using either transfectant. However, when presenting the weak agonist with B7–1 co-stimulation you get perfectly good stimulation in terms of IL-5 secretion, IL-4 secretion and thymidine incorporation, but with B7–2 co-stimulation one gets a meagre response — virtually none compared to stimulation with native peptide. The idea is that when one has a so-called 'danger' signal there is induction of B7–1, which then takes a weak agonist, which may be a cross-reactive peptide that may normally not be doing anything, and drives it to a potentially pathogenic response where you get a clonal expansion. This makes sense, because if you have viral infection you want all this cross-reactivity to be able to drive an immunological response. It may also relate to how B7–1 and B7–2 may be somewhat different in terms of what they are able to trigger.

Allen: We have taken dendritic cells and tested their ability to present our panel of ligands. All of the agonists are presented at lower concentrations. However, no non-stimulatory ligand became an agonist. Therefore, there's still some threshold for activation that we don't understand.

References

Anderson DE, Ausubel LJ, Krieger J, Höllsberg P, Freeman GJ, Hafler DA 1997 Weak peptide agonists reveal functional differences in B7–1 and B7–2 costimulation of human T cell clones. J Immunol 159:1669–1675

Hagerty DT, Evavold BD, Allen PM 1994 Regulation of the costimulator B7, not class II major histocompatibility complex, restricts the ability of murine kidney tubule cells to stimulate CD4$^+$ T cells. J Clin Invest 93:1208–1215

Jameson SC, Hogquist KA, Bevan MJ 1994 Specificity and flexibility in thymic selection. Nature 369:750–752

Liu GY, Fairchild PJ, Smith RM, Prowle JR, Kioussis D, Wraith DC 1995 Low avidity recognition of self-antigen by T cells permits escape from central tolerance. Immunity 3:407–415

Mitchison NA 1993 T cell activation states. The Immunologist 1:78–80

Viner NJ, Nelson CA, Deck B, Unanue ER 1996 Complexes generated by the binding of free peptides to class II MHC molecules are antigenically diverse compared with those generated by intracellular processing. J Immunol 156:2365–2368

Yagi J, Janeway CA 1990 Ligand thresholds at different stages of T cell development. Int Immunol 2:83–89

Tolerance and determinant hierarchy

S. C. Schneider*, H.-K. Deng¶, B. Rottinger†, S. Sharma‡, M. Stolina‡, C. Bonpane, A. Miller, S. Dubinett‡, B. Kyewski† and E. Sercarz*

Department of Microbiology and Molecular Genetics, UCLA, Los Angeles, USA, †Deutsches Krebsforschungszentrum, Heidelberg, Germany, and ‡School of Medicine, UCLA, Los Angeles, USA

Abstract. The overall T cell response to a multideterminant antigen consists of the sum of responses to a limited number of different determinants on the protein. Antigen-presenting cells (APCs) are crucial in delimiting the determinants on the protein to which a response will be mounted. This influence is apparent at two levels. First, the determinants that are generated and displayed by APCs in the thymus are pivotal in shaping the T cell repertoire that will be available for responding to antigen determinants in the periphery. Second, antigen processing affects the selection of determinants that become displayed by the various peripheral APC populations that are involved in inducing and promoting a T cell response. We have studied the effect of the display hierarchy on tolerance induction to individual determinants in transgenic mice expressing different serum levels of hen egg lysozyme. We have also analysed aspects of the processing machinery that contribute to shaping the hierarchy of determinant display on MHC class II molecules: proteolysis and reduction of disulfide bonds.

1998 Immunological tolerance. Wiley, Chichester (Novartis Foundation Symposium 215) p 54–72

The T cell response is directed to only a limited number of all theoretically possible peptide determinants of a given protein antigen. This phenomenon has been studied extensively in our laboratory by the analysis of the T cell proliferative response to overlapping synthetic peptides covering the sequence of a protein antigen after mice have been immunized with the whole protein in adjuvants (complete Freund's adjuvant; CFA). This kind of analysis (called a 'pepscan') has been applied to the study of the hierarchy of response ('determinant hierarchy') to hen egg lysozyme (HEL) as a foreign antigen in many different mouse strains (Gammon et al 1991). In each strain, only one or a few regions in the molecule give a consistent and strong recall of the response induced by whole HEL; these are called dominant determinants. Some determinants give a weaker recall that may

Present addresses: *La Jolla Institute for Allergy and Immunology, 10355 Science Center Drive, San Diego and ¶Antigen Express Inc., 1 Innovation Drive, Worcester, MA 01605, USA not be apparent in every individual mouse, the subdominant determinants. Cryptic determinants never recall the response induced by whole protein; they are however immunogenic, indicating that there is a T cell repertoire available to recognize the peptide. (Peptides that do not bind to the respective MHC class II molecules are not being considered here.)

The reasons for this determinant hierarchy can be found on two levels: the T cell repertoire and antigen processing (Sercarz et al 1993). A determinant may be subdominant or even appear cryptic because the frequency and/or affinity of T cells in the mouse is too low to allow the detection of a proliferative response. Furthermore, the specific T cells may be of a subtype that does not proliferate well, but might respond by producing cytokines (L. Soares, A. Miller & E. Sercarz, unpublished results). On the other hand, a determinant may not be processed at all or not be displayed on cell surface MHC class II molecules at sufficient levels to induce a response. Additionally, processing of some determinants may result in peptides with residues hindering the recognition by certain T cell receptors (TCRs) (Moudgil et al 1996) or alternatively lead to the removal of residues critical for TCR recognition.

As the ligand recognized by the TCR is a peptide bound to the MHC molecule, it is important to look beyond the generalization of a T cell responding to or being tolerized by an 'antigen' and take a detailed look at the individual determinants. It is obvious that only determinants that are displayed on cell surface class II molecules at sufficient levels can shape the T cell repertoire during thymic or peripheral selection or induce a response in the periphery. Here we are going to report on our experiments using HEL transgenic (Tg) mice to study the effect of different antigen levels on tolerization of individual determinants. Furthermore, we have studied the contribution of determinant display to the response hierarchy. As different APC types are responsible for the induction of tolerance and the priming or propagation of an immune response, we analysed a wide range of APC lines as well as freshly isolated APC types for potential functional differences in their processing machinery.

Determinant hierarchy: crypticity can occur on the level of display or of the T cell repertoire

We have extensively analysed the response hierarchies to HEL in H-2^d mice. Figure 1 shows the pepscan of B10.D2 mice (H-2A^d, E^d) immunized with whole HEL in CFA. There is a prominent response in the area of 106–116, the E^d-restricted dominant determinant in the H-2^d haplotype. The subdominant responses are to the determinants 11–25 (A^d) and 21–32 (A^d). In B10.GD mice, which lack the H-2E^d molecule, 11–25 becomes the dominant determinant and 21–32 remains subdominant. In none of the H-2^d strains that we have analysed could we detect a

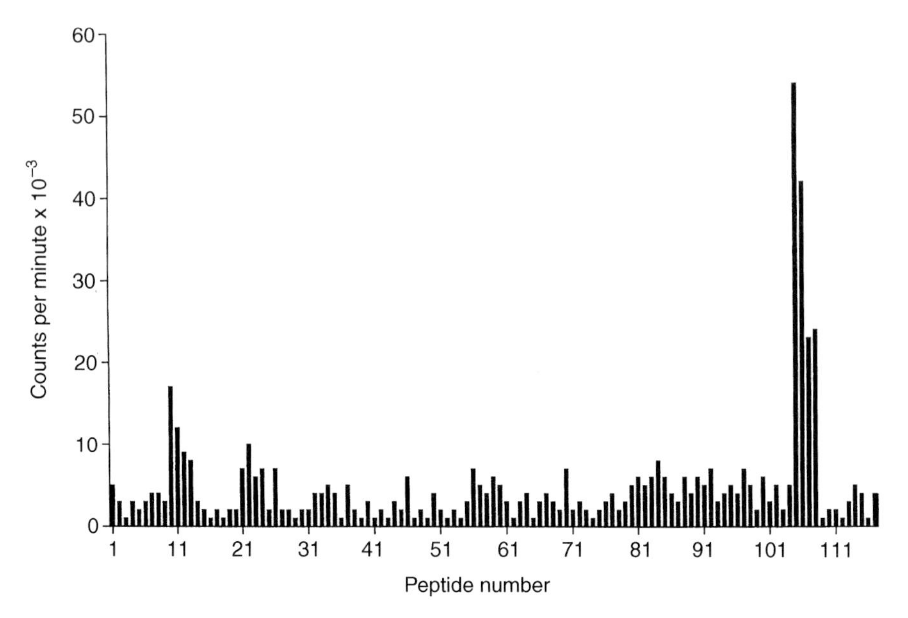

FIG. 1. Determinant hierarchy of HEL in B10.D2 mice. Mice were immunized with 7 nmoles HEL in CFA in their hind footpads. Draining lymph nodes were taken 10 days later and proliferation measured by [^{3}H]thymidine uptake to overlapping 15mer peptides covering the HEL sequence. The number of the first amino acid of each peptide used for recall is indicated.

significant proliferative response to either 1–16 or 71–85, which are therefore classified as cryptic determinants. However, we have been able to derive a 71–85/A^d-specific T cell hybridoma from HEL-immunized B10.GD mice. This T cell hybridoma responds very well to APCs provided with exogenous HEL. It thus provides an example of crypticity on the level of the T cell repertoire; the determinant is efficiently processed and displayed by APCs but the frequency or subtype of T cells seems to be too low to generate a significant proliferative response. It is, however, possible to recover these rare cells as T cell hybridomas. On the other hand, a T cell hybridoma specific for the cryptic 1–16/E^d could be derived only from peptide-immunized mice and it does not respond to whole HEL presented by a number of different APC types, including dendritic cells (DCs). It therefore seems that this region is not efficiently processed from the whole molecule by most APCs. Particularly for self-antigens, it is important to understand at which level the crypticity of a given determinant is established—crypticity as a result of inefficient display can potentially be reversed. As T cells are not tolerized to these generally insufficiently displayed determinants, their sudden display, e.g. under inflammatory conditions, could lead to autoreactive responses and potentially to autoimmune disease.

Antigen levels affect tolerance to individual determinants distinctly

Kanellopoulos and co-workers have generated transgenic mice expressing HEL under a ubiquitous promoter (Cibotti et al 1992). The special feature of these mice is that even within the same litter the HEL levels vary markedly between individuals (2 to >300 ng/ml). In a given animal, however, the variation of HEL blood levels does not exceed 30%. These mice are therefore valuable tools for studying the influence of antigen levels on tolerance induction. Cibotti et al (1992) reported that mice with HEL serum levels above 10 ng/ml selectively lose the proliferative response to the dominant determinant but responses to two subdominant determinants could still be induced and recalled with peptides. These results can be interpreted as indicating that the dominance of determinant 108–116 in non-Tg animals is a consequence of its efficient display by APCs. When HEL becomes a neo-self antigen, this determinant will be the first to reach the threshold for deletion. This notion is supported by our findings that constitutive presentation of this dominant determinant by APCs isolated from various organs of HEL-Tg mice could be detected using a specific T cell hybridoma, whereas three other determinants could not be detected using equally sensitive T cell hybridomas (data not shown).

We used these HEL-Tg mice to take a more thorough look at the effect of antigen levels on tolerization of individual determinants. Figure 2a shows that the determinant hierarchy of non-Tg littermates (BALB/c × DBA/2) differs slightly from that of B10.D2 mice in that the subdominant determinant 21–32 is not apparent and the response to 11–25 is quite small. In agreement with the results of Cibotti et al (1992), the response to the dominant determinant 106–116 is almost completely abolished in HEL-Tg mice with 20 ng/ml HEL in their serum (Fig. 2b), while the small response to 11–25 is practically not affected. Strikingly, not only is the response to the subdominant determinant 21–32 strongly enhanced in these mice, but they also mount a response to the region 49–63, to which a response has not been detected so far in any other H-2^d mouse strain. It is conceivable that the response to this novel determinant is a result of positive selection of T cells that can only occur within a very narrow concentration range. Indeed, *in vitro* experiments in other systems have demonstrated that thymocytes can be positively selected by their nominal antigen in a narrow dose range (Hogquist & Bevan 1996). Alternatively, the absence of the dominant population of 106–116-specific T cells may alter the cytokine balance in the lymph nodes during priming sufficiently to allow a more substantial number of 21–32 and 49–63-specific T cells to be induced and proliferate. In either case, 49–63 is another example of a determinant that can be generated by APCs from whole HEL, but appears cryptic because of constraints

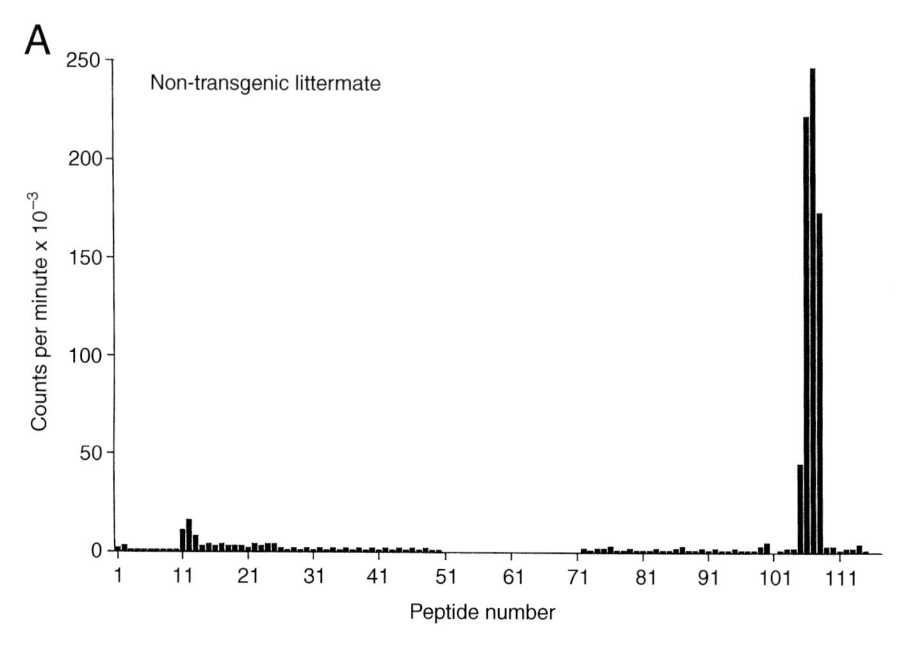
A
Non-transgenic littermate
Counts per minute x 10^{-3}
250
200
150
100
50
0
1
11
21
31
41
51
61
71
81
91
101
111
Peptide number

FIG. 2A

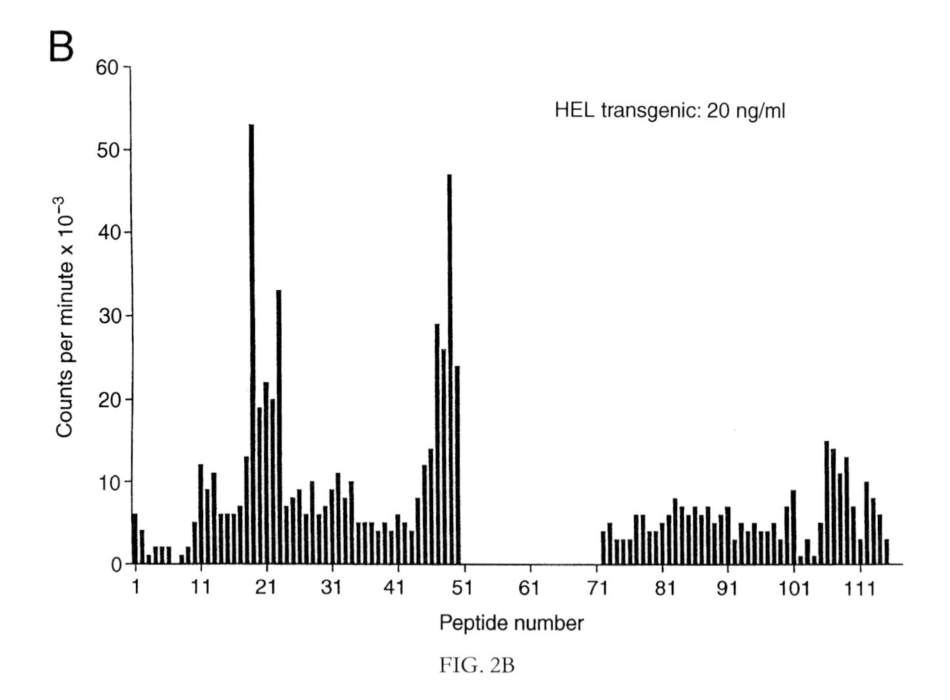
B
HEL transgenic: 20 ng/ml
Counts per minute x 10^{-3}
60
50
40
30
20
10
0
1
11
21
31
41
51
61
71
81
91
101
111
Peptide number

FIG. 2B

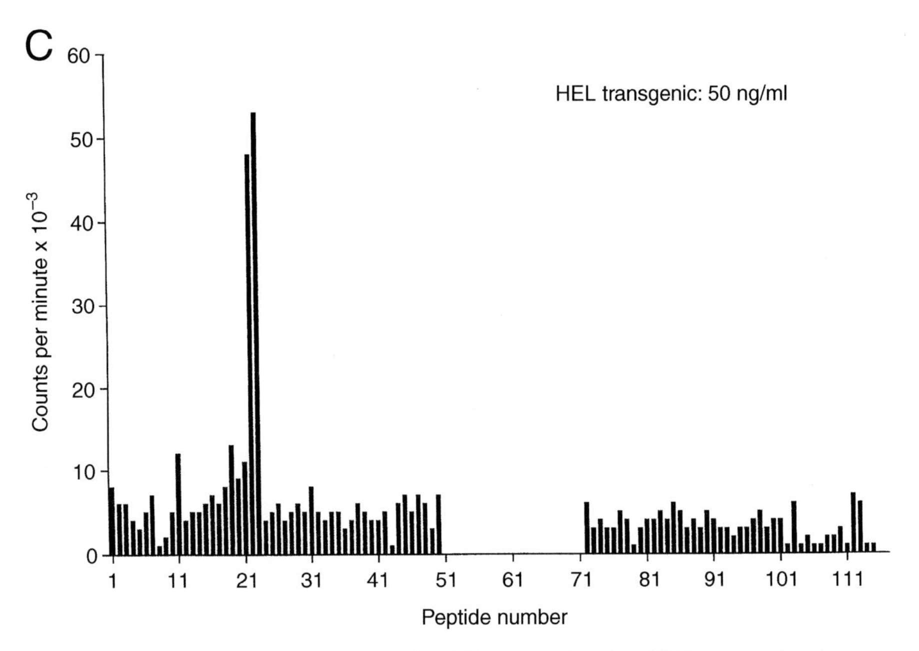

FIG. 2. Determinant hierarchy of HEL in HEL transgenic mice. HEL transgenic mice were immunized and analysed as described in the legend to Fig. 1. (A) Non-transgenic littermates; (B) HEL transgenic mice expressing 20 ng/ml HEL in their serum; (C) HEL transgenic mice expressing 50 ng/ml HEL in their serum. Peptides 52–71 were not synthesized.

in the T cell repertoire in normal mice. Interestingly, the response to 49–63 is lost again in those Tg mice in which the HEL serum levels reach 50 ng/ml (Fig. 2c). The only detectable response in these mice is directed to the determinant 21–32. Thus, the T cell repertoire and ultimately the determinant hierarchy can be profoundly influenced by small changes in the dose of self-antigens.

Antigen processing affects the display hierarchy

In order to understand the complex interactions that result in the determinant hierarchy *in vivo* it is necessary to analyse the different aspects that contribute to it separately. Thus, we studied the impact of antigen processing on the display of individual determinants on cell surface class II molecules of diverse APCs. We analysed two important aspects of antigen processing: (1) proteolysis and (2) disulfide reduction.

A putative protease cleavage site: the dibasic motif

To analyse the role of specific protease sites for the processing of a set of determinants from HEL, we generated recombinant HEL into which we introduced single amino acid exchanges using site-directed mutagenesis. Several enzymes, including cathepsin B, have the ability to cleave at a dibasic sequence. The initial endopeptidase activity is generally followed by a carboxypeptidase, leading to peptide chain cleavage and the loss of both basic residues (Bond & Butler 1987).

We confirmed this scenario by attempting to influence the generation of a well analysed determinant: 48–61/A^k, as recognized by the T cell hybridoma 3A9 (Nelson et al 1996). This particular T cell hybridoma requires arginine at position 61 in order to be stimulated. We made a recombinant HEL in which we changed position 62 from tryptophan to arginine or lysine, in each case creating a dibasic motif. If proteases that are sensitive to dibasic motifs are involved in the processing of HEL, the introduction of a dibasic motif at position 61/62 should lead to the loss of R61 in the processing of the mutant HEL. In agreement with this hypothesis, stimulation of the T cell hybridoma 3A9 by either mutant HEL (W62K or W62R) was reduced as compared to its stimulation by HEL (H.-K. Deng, S. C. Schneider, J. Ohmen, J. Guo, L. Fosdick, B. Gladstone, H. Bang, D. Yi-Ti Tsai, A. Miller, B. S. Kim & E. Sercarz, unpublished results).

One of the two naturally occurring dibasic motifs in the sequence of HEL (K96K97) is adjacent to the determinant 87–96, which is cryptic in the H-2^k haplotype. Representative of this situation, the T cell hybridoma AOIT13.1 recognizes the synthetic peptide 87–96 but does not respond to whole HEL. As K96 is required for the stimulation of this T cell hybridoma by the synthetic peptide, we hypothesized that the loss of K96 as a result of the cleavage at the dibasic motif in the processing of HEL explains the T cell hybridoma's unresponsiveness to HEL. To test this hypothesis, we destroyed the dibasic motif by making a mutant HEL in which K97 was changed to L (K97L). This should prevent the loss of K96 as a result of the cleavage of the dibasic motif. Indeed, AOIT13.1 was stimulated efficiently by K97L but not by any unrelated mutant HEL (A. Ametani, H.-K. Deng & E. Sercarz, unpublished results).

Having established the capacity of dibasic motifs to alter the processing of HEL, we set out to make a subdominant determinant (21–32/A^d) more available by flanking it with a dibasic motif. As position 33 is K, this was achieved by mutating F34 to R or K. As shown in Fig. 3, the B lymphoma line A20 stimulates the 21–32/A^d-specific T cell hybridoma about 40-fold more efficiently when provided with the mutant F34R than with HEL. Introducing the amino acid exchange F34R did not affect the recognition of a synthetic peptide 21–35 by the T cell hybridoma (data not shown), indicating that the introduction of the dibasic motif alters the processing of this determinant, rather than T cell recognition. Although the exact boundaries of the set of naturally processed peptides are not

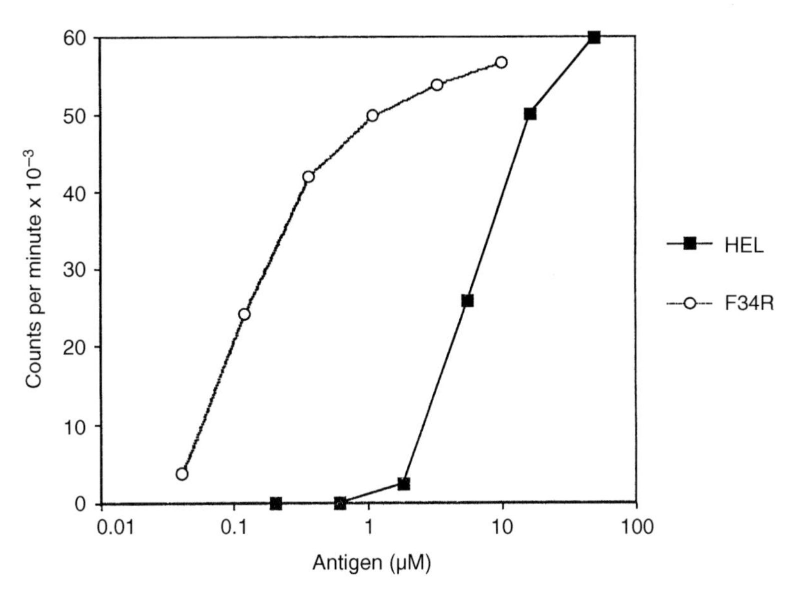

FIG. 3. Introduction of a dibasic motif adjacent to the subdominant determinant 21–32 enhances its presentation by A20. Presentation of HEL and its variant F34R by the B lymphoma line A20 to the 21–32/A^d-specific T cell hybridoma 6F7. IL-2 secretion was measured in a HT-2 assay.

known for this determinant we analysed a set of nested synthetic peptides for their capacities to stimulate the T cell hybridoma (data not shown). Fixing the C-terminus of the peptide at 32 (the likely result of the introduced mutation) only slightly decreased the stimulatory capacity of a peptide starting at position 20. Varying the N-terminus of peptides ending at position 32 did not significantly alter T cell recognition unless the peptide was as short as 23–32. We cannot exclude that the altered processing leads to the generation of more stimulatory peptide species, but the results with synthetic peptides suggest that the enhancement of T cell stimulation is due to an increase in the total amount of the determinant generated by the APCs from F34R than from HEL.

Functional differences in processing between APC types

In spite of the *in vitro* results, immunization of BALB/c mice with the mutant F34R did not enhance the proliferative response to the determinant 21–32 as compared with HEL immunization (data not shown). While this could be due to a low frequency of specific T cells in these mice, it was also conceivable that not all

APCs that are involved in the induction and propagation of an immune response process the mutant equally well. We therefore analysed different types of freshly isolated APCs from various organs for their capacities to process F34R vs. HEL for presentation to the 21–32/A^d-specific T cell hybridoma.

As shown in Table 1, two B lymphoma lines generated the determinant 21–32 far more efficiently from F34R than from HEL (20–40-fold). However, freshly isolated DCs from various organs as well as splenic macrophages displayed only a slight enhancement of presentation from F34R vs. HEL. As DCs are important for the priming of a T cell response, their failure to present 21–32 more efficiently from F34R than from HEL might be sufficient to explain the lack of enhancement of the proliferative response, even in the presence of B cells which might be capable of processing F34R more efficiently. Alternatively, the more efficient presentation of 21–32 from F34R might be an effect unique to B lymphomas that is not exhibited by fresh B cells. In this case, no enhancement of the response to 21–32 by immunization with F34R *in vivo* would be expected. However, this would mean that the processing machinery of B lymphoma lines differs sufficiently from that of fresh B cells to generate a drastic, determinant-specific difference in display. Much of the current knowledge about MHC class II trafficking and the localization of proteolytic enzymes in APCs is based on the analysis of B lymphoma lines. It would thus be important to be aware of significant functional

TABLE 1 Introduction of a dibasic motif adjacent to the subdominant determinant 21–32 enhances its presentation by B lymphomas but not by other APC types

Antigen-presenting cell	*HEL index*[a]
B lymphomas	
A20	38
LB27.4	24
Fresh peripheral APCs	
Dendritic cells	
Lymph node	3
Splenic	4
Langerhans' cells	3
Splenic macrophages	2
Fresh thymic APCs	
Dendritic cells	3
Epithelial cells	4

[a]HEL index: x-fold HEL required for half-maximal stimulation of the HEL21-32/A^d-specific T cell hybridoma than F34R.

differences between such lines and fresh B cells. Furthermore, B cells might handle F34R differently depending on their activation state and route of uptake.

All APCs listed in Table 1 were also tested for their ability to present HEL and F34R to T cell hybridomas specific for three other determinants. We could not detect any significant differences in their efficiencies to present each determinant from HEL as opposed to F34R (data not shown). Thus, the constitutive processing machinery of APCs of various lineages isolated from various organs seems sufficiently similar.

Importantly, no difference in the display hierarchy could be detected between thymic and peripheral APCs. Thymic epithelial cells (freshly isolated as thymic nurse cells) and thymic DCs presented HEL and F34R to the 21–32/A^d-specific T cell hybridoma with comparable efficiencies (Table 1). However, a cortical thymic epithelial cell line did not generate a determinant from HEL, but was efficient at stimulating the T hybridoma when provided with F34R (H.-K. Deng, S. C. Schneider, J. Ohmen, J. Guo, L. Fosdick, S. Gladstone, H. Bang, D. Yi-Ti Tsai, A. Miller, B. S. Kim & E. Sercarz, unpublished work). It is important to point out that the fresh epithelial cells express high levels of class II and are very efficient APCs, whereas the epithelial cell line requires induction of class II expression with interferon (IFN)-γ to acquire the ability to present antigen. In macrophages, IFN-γ has been shown to cause a selective induction of the lysosomal proteases cathepsin B and L (Lah et al 1995). It is conceivable that in epithelial cells, IFN-γ affects not only class II expression but also some other step of the processing pathway that leads to the observed difference in generating the determinant 21–32.

The role of disulfide cleavage in antigen processing

Besides proteolytic cleavage of antigen, disulfide bond reduction and protein unfolding are important events during antigen processing. Denatured and reduced proteins probably bind to class II molecules intracellularly (Sercarz et al 1986, Sette et al 1989) and are proteolytically trimmed further before being presented at the cell surface (Nelson et al 1997). Furthermore, reduction of disulfide bonds may be required to render proteins susceptible to proteolytic attack. Most notably, HEL, which is held in a compact conformation by four intracellular disulfide bonds, is resistant to proteolysis by purified proteases as well as endosomal or lysosomal fractions *in vitro* unless it is reduced (Collins et al 1991, Van Noort & Jacobs 1994) or nicked (Benavida et al 1969).

The intracellular compartment responsible for reduction of endocytosed molecules is still controversial (Jensen 1995). Several groups have reported reduction of disulfide-linked marker compounds in either early or late endosomes, lysosomes or even Golgi in different cell types. It is therefore unknown whether reduction is an early event that is completed before the first

TABLE 2 The provision of reduced HEL affects the presentation of four determinants differentially

	X-fold HEL needed over RCM-HEL for half-maximal stimulation of T hybridomas specific for:			
Antigen-presenting cell	*11–25/A^d*	*21–32/A^d*	*71–85/A^d*	*106–120/E^d*
Dendritic cells	20–50	3–10	<<	0.05–1
Macrophages	3–4	1–3	<<	0.08–0.18
B lymphomas	2–4	1–5	<<	0.1
Peptide stimulation	No C	A>>C	AA>>CC	A<C

<<, no stimulation of the T hybridoma by RCM-HEL but regular stimulation by HEL peptide stimulation; A>>C, better stimulation by a synthetic peptide in which C has been changed to A.
RCM-HEL, reduced carboxymethylated HEL.
Dendritic cells: comparable results for dendritic cells freshly isolated from thymus, spleen, lymph node and from bone marrow cultured with GM-CSF and IL-4.
Macrophages were freshly isolated from spleen.
B lymphomas: A20 and LB27.4.

proteases have access to the endocytosed molecules or whether it is a slow process, intertwined with the action of proteases. In the latter case, the reduction state of the antigen when endocytosed should affect the pattern of proteolytic cleavage and might alter the set of determinants generated from the protein antigen. We therefore compared the efficiency of the generation of four determinants from native HEL with that from a completely reduced version, carboxymethylated HEL (RCM-HEL) and 3SS-HEL, in which only the C6–C127 bond is reduced and carboxymethylated.

As shown in Table 2, providing reduced HEL enhances the presentation of the determinant 11–25/A^d by all APCs, but most dramatically (20- to 50-fold) by DCs from a variety of sources. Similarly, presentation of 21–32/A^d from reduced HEL is more efficient than from HEL for all APCs; again, the effect is slightly more marked for DCs. The cysteine does not contribute significantly to T cell recognition of this determinant. In both cases, reduction of C6–C127 is sufficient for the effect (data not shown). Two other determinants, 71–85/A^d and 106–116/E^d, are presented less efficiently from the reduced versions of HEL than from native HEL by all APCs (Table 2). However, both determinants are presented better from 3SS-HEL than from RCM-HEL, which is likely due to steric hindrance by the bulky carboxymethyl groups on the cysteines in the determinants. And while their presentation from 3SS-HEL is still worse than from native HEL, it is much better in DCs than in B cells (data not shown).

Our results are consistent with the idea that the reduction of disulfide bonds of endocytosed HEL is a process that overlaps with the action of proteases on the molecule. Furthermore, APC types differ in their potential to reduce endocytosed antigens. DCs generate the determinant 11–25, and to a lesser degree 21–32, much more efficiently when HEL is provided in its reduced form than in its native form. Thus, DCs might be less efficient at reducing endocytosed antigen than B cells or macrophages. Furthermore, reduction of the disulfide C6–C127 seems to be sufficient to render HEL sensitive to proteolytic attack.

Summary

As T cell tolerance is achieved to individual determinants, it is important to analyse the parameters that determine their generation and display on APCs. We showed that the thresholds for tolerization of T cells specific for individual determinants of a self-antigen can be different. This might be the consequence of inherent differences in the repertoires of the specific T cells or of differences in the efficiency of processing. Furthermore, the total immune response to an antigen depends on the determinants that are displayed by the diverse APC types involved in inducing and expanding a T cell response. Our system allowed us to detect functional differences in processing between APC lineages; both in their efficiency to reduce endocytosed antigen and in their proteolytic machinery. However, no differences were detected between thymic and peripheral APCs; thus, under normal circumstances the maintenance of self-tolerance to particular determinants seems to be ensured. Any change in the processing machinery due to external influences, e.g. inflammatory cytokines, could threaten this equilibrium and potentially cause an autoimmune reaction.

References

Bonavida B, Miller A, Sercarz EE 1969 Structural basis for immune recognition of lysozymes. I. Effect of cyanogen bromide on hen egg-white lysozyme. Biochemistry 8:968–979

Bond J, Butler P 1987 Intracellular proteases. Annu Rev Biochem 56:333–364

Cibotti R, Kanellopoulos J, Cabaniols J-P et al 1992 Tolerance to a self-protein involves its immunodominant but does not involve its subdominant determinants. Proc Natl Acad Sci USA 89:416–420

Collins DS, Unanue ER, Harding CV 1991 Reduction of disulfide bonds within lysosomes is a key step in antigen processing. J Immunol 147:4054–4059

Gammon G, Geysen HM, Apple RJ et al 1991 T cell determinant structure: cores and determinant envelopes in three mouse MHC haplotypes. J Exp Med 173:609–617

Hogquist KA, Bevan MJ 1996 The nature of the peptide/MHC ligand involved in positive selection. Semin Immunol 8:63–68

Jensen P 1995 Antigen unfolding and disulfide reduction in antigen presenting cells. Semin Immunol 7:347–353

Lah T, Hawley M, Rock K, Goldberg A 1995 Gamma-interferon causes a selective induction of the lysosomal proteases, cathepsins B and L, in macrophages. FEBS Lett 363:85–89

Moudgil KD, Grewal IS, Jensen PE, Sercarz EE 1996 Unresponsiveness to a self-peptide of mouse lysozyme owing to hindrance of TCR-MHC complex/peptide interaction caused by flanking epitopic residues. J Exp Med 183:535–546
Nelson C, Viner J, Unanue E 1996 Appreciating the complexity of MHC class II peptide binding: lysozyme peptide and I-A^k. Immunol Rev 151:81–105
Nelson CA, Vidavsky I, Viner NJ, Gross ML, Unanue ER 1997 Amino-terminal trimming of peptides for presentation on major histocompatibility complex class II molecules. Proc Natl Acad Sci USA 94:628–633
Sercarz EE, Wilbur S, Sadegh-Nasseri S et al 1986 The molecular context of a determinant influences its dominant expression in a T cell response hierarchy through 'fine-processing'. In: Progress in Immunology IV: sixth International Congress of Immunology, Academic Press, Orlando, p 227–237
Sercarz EE, Lehmann PV, Ametani A, Benichou G, Miller A, Moudgil K 1993 Dominance and crypticity of T cell antigenic determinants. Annu Rev Immunol 11:729–766
Sette A, Adorini L, Colon SM, Buus S, Grey HM 1989 Capacity of intact proteins to bind to MHC class II molecules. J Immunol 143:1265–1267
Van Noort J, Jacobs M 1994 Cathepsin D, but not Cathepsin B, releases T cell stimulatory fragments from lysozyme that are functional in the context of multiple murine class II MHC molecules. Eur J Immunol 24:2175–2180

DISCUSSION

Mitchison: What do you think the dibasic substitution is doing?

Schneider: It is a known target site for several endopeptidases, but it is unclear which enzyme is involved in our case. The dibasic sequence is a common proteolytic motif, and it is likely that the creation of a dibasic site will change the way the molecule is processed.

Mitchison: I know that the study of individual proteases, for example protease E, is a very active field. I believe you're collaborating with Benjamin Chain (UCL, London) on that subject. Are there differences in the expression of different cathepsins among the proteases between dendritic cells (DCs) and B cells? I thought that cathepsin E is likely to be an important protease in B cells but not in DCs.

Schneider: Cathepsin E is expressed in DCs and B cells but not in monocytes/macrophages, whereas cathepsins B and D are ubiquitous. Also, the intracellular localization differs between cathepsins: cathepsin E is localized to early endosomes and the endoplasmic reticulum, cathepsin D is limited to lysosomes and late endocytic compartments, and cathepsin B is expressed in most endocytic compartments. We don't know which protease is responsible for the effect we see. We have begun preliminary experiments with members of the Chain lab in which we are studying the *in vitro* digestion of the various mutants with the individual cathepsins. Looking at cathepsin D and E, we do get different patterns of digestion between native HEL, F34R and the W62K or R, but the K and R look the same.

Allen: One other interpretation may be as follows. Daved Fremont has solved the crystal structure of HEL(48–62)/I-A^k. The 60 residue is the last anchor residue, and 61 and 62 are sticking out of the groove. The question is, if these additional residues are present in the naturally processed form of the peptide, what is their function? It is possible that they could be contacting some part of the T cell receptor. I think you should be cautious in your interpretation: perhaps what you are doing here is affecting T cell recognition, because you are changing these residues. Going from a tryptophan to a lysine is a huge change, and instead of the processing effects, you might be affecting T cell recognition.

Schneider: We cannot exclude some influence of T cell recognition on the effect we are observing with 48–61, although our peptide controls indicate that changes in T cell recognition cannot account for the total shift in the dose–response curve of the T hybridoma 3A9. With 21–35, the profiles of synthetic peptides with or without the mutation are identical, so I'm more confident for this determinant that we are not drastically altering T cell recognition. Specifically for the determinant 48–61, not only are we introducing a dibasic motif, but we are also destroying a dihydrophobic one, either of which might alter the processing pattern.

Hämmerling: You mentioned the possibility that trafficking of MHC class II molecules might be different in the cells you've been looking at. There is actually good evidence that trafficking in freshly isolated (and therefore immature) DCs is different from that in mature DCs, where class II molecules are found in a different compartment (Pierre et al 1997). In the early DCs they are mostly present in the lysosomal compartment, but at the mature stage they are more commonly found in the compartment which Ira Melmann calls the CIIV, a multilaminar compartment found early in the endocytic pathway. In A20 cells, the class II-positive compartments are mainly the CIIV, so if you are intending to generalize and extend your findings to other cells, you may be comparing apples with pears. In addition to the distribution of class II it will also be important to determine the distribution of cathepsins in these compartments.

Shevach: Isn't it a little unfair to compare the malignant B lymphoma cells with the freshly isolated and cultured DCs?

Schneider: Definitely. We have to study fresh B cells to determine whether the functional difference in processing that we are detecting is between B lymphomas and fresh B cells or between B cells and other APC lineages.

Shevach: Have you done these experiments, for example, with CD40 ligand-activated B cells, which are powerful APCs?

Schneider: We intend to do these next.

Abbas: Can I go back to the first part of your talk in which you discussed cryptic determinants. If I understand it correctly, those soluble HEL-expressing mice are not tolerant to some of the subdominant epitopes: is that correct?

Schneider: Yes, there is a direct relationship between the serum levels of antigen and the level of tolerance induced to individual determinants.

Abbas: If you take those mice and either immunize them with HEL or just examine them, do they make antibody? Is there any evidence for autoimmunity in those animals? In other words, if they have soluble HEL and they're not tolerant to subdominant epitopes, their B cells should present the HEL and trigger appropriate reactions.

Schneider: That is a good point. The HEL transgenic system is very useful for dissecting response hierarchies to individual determinants, but in terms of interpreting functional tolerance, you need to look at a model where not being tolerant is bad for the mouse.

Mitchison: Cibotti et al (1992) have published quite substantial studies on these mice already.

Allen: We did a study a couple of years ago using the Goodnow/Basten HEL transgenic mice. We could detect HEL reactive cells and antibodies in these mice, but we never saw any disease at all. I think there's still quite a lot of variation in the concentration effects of different neo-self antigens.

Shevach: Did you get proliferative T cells without priming the mice?

Allen: No. If we immunized with HEL/CFA with pertussis, then we got a robust T cell response. Our results were a little different in that we saw the whole spectrum of T cells. We saw that 46–61 was still present, and we didn't see that this was eliminated.

Abbas: I don't understand how the repertoire size can explain crypticity. Eli Sercarz's definition of crypticity is that you get the response to a peptide epitope when you give the peptide epitope itself, but not when you give it as a part of the intact protein. Whether you give it as an epitope or as an intact protein, the frequency of reactive T cells is exactly the same. I don't understand how the repertoire accounts for cryptic epitopes.

Schneider: But when you immunize a mouse with the whole protein, several determinants will be presented by the APC and will be able to induce T cell responses. This can result in competition between T cells with different specificities, for example on the level of cytokine competition. Therefore, a response to a particular determinant that can be induced with a synthetic peptide might not be revealed when you immunize with the whole protein.

Mason: It is quite possible for animals to have autoreactive cells that don't cause any disease. In this case you can show that they are present by immunizing the animal with the relevant antigen in CFA.

Mitchison: I don't think those Kanellopoulos mice developed antibodies spontaneously.

Abbas: Are you talking here about relatively anatomically sequestered antigens?

Mason: There is good evidence that insulin is an important autoantigen in insulin-dependent diabetes. It is not a sequestered antigen but its concentration in the blood is very low.

Allen: If you push the immune system hard enough you can find T cells against many things. However, if you can only find reactivity to an antigen when you use CFA or pertussis, its function is probably irrelevant.

Schneider: As I said earlier, you need to look at a disease model to say something about the functional effect of tolerance, because just the presence of a certain response really doesn't tell you whether this response would be detrimental to the individual or not.

Abbas: Diane Mathis and others have talked about this checkpoint between autoreactivity and disease development (Katz et al 1993). I don't know if anybody understands what the boundary is.

Wraith: The impression I got from what Susanne Schneider was saying was that she was studying the difference between dominance and subdominance, not crypticity.

Susanne, have you actually looked at those epitopes, found out what the naturally processed epitope is, and then measured the affinity of those epitopes for MHC?

Schneider: No, not directly. But with regard to your first point about the difference between subdominance and crypticity, the whole thing is of course a continuum. I pointed out, for example, the determinant 1–16, which in our hands never gets presented to the specific T cell hybridoma. Hal Drakesmith in Peter Beverley's lab has used this hybridoma and has got DCs to present this determinant from whole HEL when the DCs are treated with IL-6. This is a very interesting finding, giving an example of a cytokine actually inducing the processing and display of an otherwise cryptic epitope. Hal's finding suggests that IL-6 changes the processing in DCs sufficiently to permit the generation of the 1–16 determinant. This argues against the idea that the hybridoma 1–16 is just a type B hybridoma as described by Nelson et al (1996). These hybridomas recognize a synthetic peptide, but the same peptide as processed from whole HEL is not recognized.

Mitchison: That's a very interesting finding, as you say, because it links differences in antigen presentation to the level of inflammation. Could you put that in a wider context? Is there anything to suggest in studies on infection and immunity, that this is a significant effect in the course of the normal response to pathogens? Is there any background literature on inflammatory-type immunizations generating different cryptic or non-cryptic epitopes?

Schneider: As far as I know, these are the first experimental data showing up-regulation of a cryptic epitope by a cytokine. There has been no report of IL-6 influencing processing to date.

Shevach: Once you invoke the idea of a cytokine influence in presentation, you bring into play a variety of inflammatory agents that induce cytokines. For example, exposure of macrophages to bacterial DNA induces the production of IL-6, IL-12 and IL-1 (Pisetsky 1996).

Hämmerling: Has this been controlled for class II expression? In certain cells treatment with IL-6 increases class II MHC expression. Therefore, the observed effect may have nothing to do with processing, but just the presence of more class II.

Schneider: Class II expression was not affected, neither was the presentation of other determinants. This is a determinant-specific effect.

Hafler: But this is an interesting question: does one ever see the amount of class II expressed as a regulatory step? It doesn't seem to be one of the great limiting steps of antigen presentation.

Hämmerling: I think the amount of expressed class II is often important. For example, uninduced macrophages express very little class II. However, after induction of class II expression they become effective APCs.

Allen: The worrying thing about the T cell repertoire is that there is a self lysozyme molecule — there's an imprint of self-tolerance imposed upon all of this. There are a lot of amino acids that are completely conserved. This would answer some of Abul's questions: you could have some repertoire changes because you're deleting those cells, because there are self molecules.

Schneider: The hierarchy of response pattern to self (mouse) and foreign (chicken) lysozymes was found to be quite different. This has been studied by Moudgil & Sercarz (1993).

Allen: But that's when you immunize. We don't know what's been deleted during thymic development.

Schneider: But mouse lysozyme is constant between transgenic and non-transgenic mice.

Allen: That could affect the repertoire.

Schneider: Absolutely. This is indicated by Kamal Moudgil's work. Despite a limited sequence homology between mouse lysozyme and HEL, he finds that the boundaries of the regions that are dominant within HEL almost match those of the cryptic determinants in mouse lysozyme. These results suggest that the anti-cryptic (mouse lysozyme) T cell repertoire might have been preferentially recruited by HEL. Kamal believes that mouse lysozyme shapes and focuses the T cell response to HEL.

Hafler: I would like to raise something that we've been grappling with in human immunology. We've been looking at T cell reactivities to a number of antigens. With myelin basic protein, proteolipid protein and tetanus toxoid, which are presumably sequestered antigens or foreign antigens, we get very highly reactive T cell clones, proliferating to 50–100 000 counts. Using identical techniques examining collagen type II and insulin, which are presumably omnipresent antigens, we can get T cell clones of reasonably high frequency, but they never have high thymidine incorporation. Why is that the case? Is it just because of high antigen concentrations which lead to clonal deletion? Could these cells still potentially be pathogenic, even though they had very low amounts of thymidine incorporation?

Wraith: To address that question it might be relevant to look at the NOD mouse. It is true that its response to insulin is relatively weak compared with ovalbumin, for instance. However, you can use a peptide from insulin to prevent disease. So there is evidence that despite the fact that the response is weak, the cells may be involved in pathogenicity since they can also be involved in tolerogenicity.

Abbas: The minute you involve bystander suppression, you lose the idea of these cells being involved in pathogenicity.

Wraith: Yes, but we can't exclude insulin being a target antigen because cells with specificity for insulin mediate bystander suppression.

Lechler: But you cannot use your tolerance data in favour of insulin-reactive T cells being pathogenic, because insulin-specific T cells that you render tolerant by some form of mucosal antigen administration could very well suppress T cells autoreactive with another self protein presented by the same APCs that have emigrated from the pancreas.

Hafler: Even in the role of bystander suppression they are very mildly reactive cells.

Abbas: One of the problems with some of the work I've seen with human T cell clones is not that they proliferate weakly or have low stimulation indices, but that the kinds of antigen concentrations needed to elicit any response are orders of magnitude higher than can possibly be accomplished *in vivo*.

Hafler: No, on two accounts. First, we can get very high thymidine incorporations on 1–10 mg of peptide. Second, if you look at brain myelin basic protein, for instance, there are certainly high enough concentrations of protein at the site of the organ to do whatever you wish to have done, because they're autoantigens.

References

Cibotti R, Kanellopoulos, Cabaniols JP et al 1992 Tolerance to a self-protein involves its immunodominant but does not involve its subdominant determinants. Proc Natl Acad Sci USA 89:416–420

Katz JD, Wang B, Haskins K, Benoist C, Mathis D 1993 Following a diabetogenic T cell from genesis through pathogenesis. Cell 74:1089–1100

Moudgil KD, Sercarz EE 1993 Dominant determinants in hen egg white lysozyme correspond to the cryptic determinants within its self-homologue, mouse lysozyme: implications in shaping the T cell repertoire and autoimmunity. J Exp Med 178:2131–2138

Nelson CA, Viner NJ, Unanue ER 1996 Appreciating the complexity of MHC class II peptide binding: lysozyme peptide and I-A^k. Immunol Rev 151:81–105

Pierre P, Turley SJ, Gatti E et al 1997 Developmental regulation of MHC class II transport in mouse dendritic cells. Nature 388:787–792

Pisetsky DS 1996 Immune activation by bacterial DNA: a new genetic code. Immunity 5:303–310

Molecular genetic studies in lymphocyte apoptosis and human autoimmunity

David A. Martin, Behazine Combadiere, Felicita Hornung, Di Jiang, Hugh McFarland, Richard Siegel, Carol Trageser, Jin Wang, Lixin Zheng and Michael J. Lenardo

Laboratory of Immunology, National Institute of Allergy and Infectious Diseases, National Institutes of Health, Bethesda, MD 20892, USA

Abstract. Using a genetic approach, we have studied the molecular basis of human autoimmunity with special emphasis on a disease that is due to defective lymphocyte apoptosis. Recently, we and our collaborators have found that the autoimmune/lymphoproliferative syndrome (ALPS), an inherited disease of children comprising marked lymphoid hyperplasia and autoimmune manifestations, is due to abnormalities in the CD95 gene that cause defective lymphocyte apoptosis. Our recent investigations have shown that the mutations in most families with ALPS cause either global or local changes in the structure of a cytoplasmic portion of the molecule called the 'death domain'. These death domain alterations impair binding of the adapter protein FADD/MORT1 and result in a failure to activate apoptotic caspases after CD95 (Fas/APO-1) cross-linking. Mutations in apoptotic caspases may also contribute to the pathogenesis of ALPS in individuals that have no CD95 gene mutations.

1998 Immunological tolerance. Wiley, Chichester (Novartis Foundation Symposium 215) p 73–87

Recently, we have approached the study of immunological tolerance by investigating the molecular genetics of the human disease autoimmune/lymphoproliferative syndrome (ALPS) in which defective lymphocyte apoptosis causes lymphoid hyperplasia and autoimmunity (Fisher et al 1995, Rieux-Laucat et al 1995). Mechanisms of immunological tolerance may be most clearly perceived by investigating autoimmune conditions in which such mechanisms fail. Furthermore, molecular biological and genetic techniques permit the delineation of tolerance mechanisms in molecular terms (Kang et al 1992, Fisher et al 1995). A genetic component is prominent in autoimmune diseases (Fig. 1). Various investigations reveal that specific gene alleles contribute to the incidence of autoimmune conditions in humans and in experimental animals (see Todd 1995 and references therein). Our studies have focused on the role of lymphocyte

Influences of genetics

- Twin studies
- Major histocompatibility locus gene association
- Multiple gene associations – 'susceptibility loci'
- Inbred strains of 'susceptible' mice and associations with certain human racial or ethnic subgroups
- Single gene knockouts in mice lead to disease

Influences of environment

- Associations with infections
- Tissue damage
- Miscellaneous: drugs, UV light, hormones, etc.

FIG. 1. Molecular genetics of autoimmune diseases. A variety of different lines of investigation have shown that both genetic and environmental influences have a strong effect on the development of autoimmune diseases.

apoptosis in tolerance because of our discovery that mature peripheral T cells that are activated and cycling in interleukin (IL)-2 will undergo apoptosis in a clonotypic manner when they are triggered through the T cell receptor (TCR) (Lenardo 1991, Boehme & Lenardo 1993) (Fig. 2). The immunological features of this effect suggests that it serves as a vital negative-feedback regulatory mechanism. We have called this mechanism propriocidal regulation because of its resemblance to proprioceptive negative-feedback mechanisms in the central nervous system (Lenardo 1991, Critchfield et al 1995).

Recent investigations of the *lpr* mouse have provided evidence for the hypothesis that mature lymphocyte apoptosis plays an important role in tolerance in the peripheral immune system (Critchfield et al 1995). When homozygous, *lpr* is a 'loss of function' mutation that causes lymphoid hyperplasia and the development of autoimmune disease on certain genetic backgrounds in mice (Fig. 3) (Cohen & Eisenberg 1991). The molecular identification of the CD95/Fas/Apo-1 receptor, a member of the tumour necrosis factor (TNF) receptor family, as the gene locus of the *lpr* mutation, solidified the link between apoptosis and mature peripheral lymphocyte homeostasis and tolerance (Nagata 1994, Nagata & Golstein 1995). However, the abrogation of CD95 function in T cells only partly prevented the death response to TCR engagement. The balance of TCR-induced death could be blocked with an antisera that inhibited TNF (Zheng

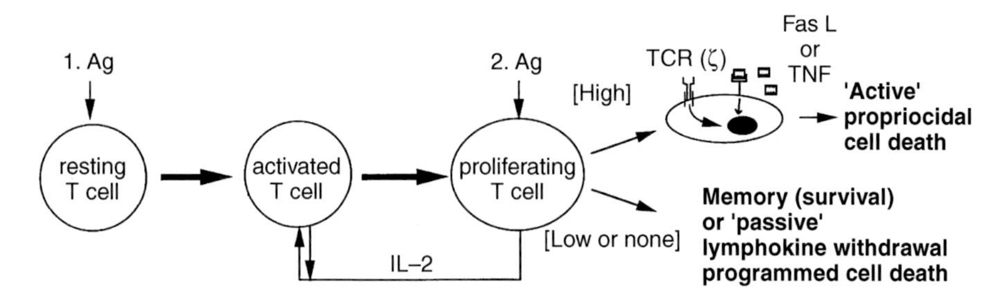

FIG. 2. A conceptual scheme of T lymphocyte apoptosis by the propriocidal and lymphokine withdrawal mechanisms (adapted from Lenardo 1996). Propriocidal or 'active' death occurs when IL-2 drives an excessive number of T cells into the cell cycle which promotes the susceptibility to apoptosis when further high concentrations of antigen (Ag) are encountered. This type of death is due to the activation of the CD95 and TNF molecular pathways of apoptosis. The principal purpose of propriocidal regulation is to provide negative feedback control over T cell expansion when persistent or recurrent antigen is encountered. By contrast, the absence of further antigen stimulation causes a decline in growth cytokines that leads to the 'passive' or lymphokine withdrawal pathway of apoptosis. The principal purpose of lymphokine withdrawal apoptosis is to provide homeostatic control and re-establish the basal number of T cells after the disappearance of antigen. Although not understood in molecular terms, some T cells may acquire the ability to escape these death pathways and survive as 'memory' cells.

et al 1995). Thus, there are at least two pathways to mature T lymphocyte apoptosis. These pathways may not be redundant since *in vitro* assays suggest that $CD4^+$ T cells typically undergo deletion by CD95 whereas $CD8^+$ T cells may typically undergo deletion by TNF. Recently, we have preliminary evidence that there is a third mechanism that may also mediate TCR-induced apoptosis (F. Hornung & M. J. Lenardo, unpublished observations).

In collaboration with Drs Stephen Straus, Warren Strober and Jennifer Puck and their associates, we have studied a group of children suffering from lymphoid hyperplasia and autoimmunity. These children had several common disease traits, including: (1) marked enlargement of the secondary lymphoid organs due to the accumulation of non-malignant lymphocytes; (2) an expanded population of $\alpha\beta$ T cells that lack the CD4 and CD8 co-receptors; and (3) evidence for autoimmunity that was typically due to anti-blood cell and anti-platelet antibodies (Sneller et al 1992). To these common features, we also consistently found a striking defect in the propriocidal mechanism of TCR-induced apoptosis for T cells. Furthermore, defective CD95-induced apoptosis caused an overaccumulation of B cells. These findings allowed the children's disorders to be classified as a common syndrome that was given the name autoimmune/lymphoproliferative syndrome (ALPS) (Fisher et al 1995). To date, in ALPS

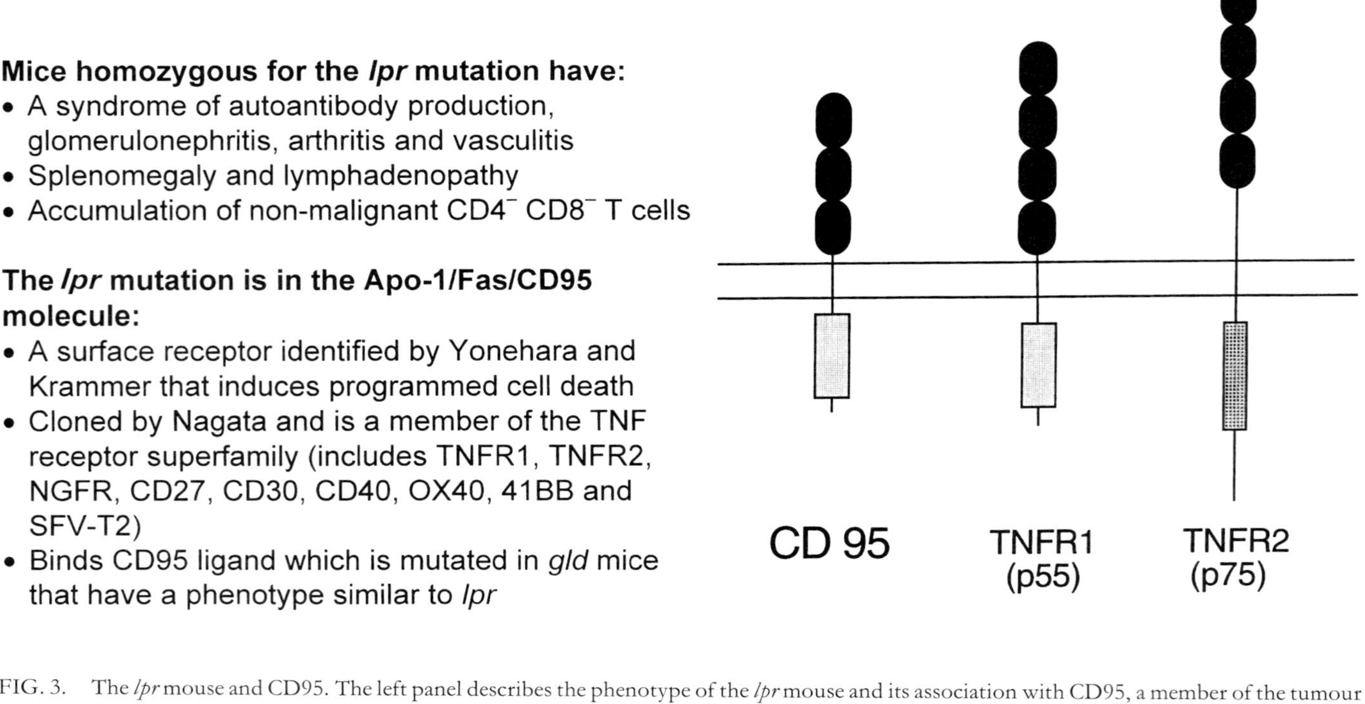

FIG. 3. The *lpr* mouse and CD95. The left panel describes the phenotype of the *lpr* mouse and its association with CD95, a member of the tumour necrosis factor receptor (TNFR) superfamily. The right panel schematically illustrates CD95 and both types of TNFR. The horizontal double line shows the cell membrane and the region above shows the cysteine-rich domains (solid black ovals) that project into the extracellular space and are important for ligand binding. The intracellular region of CD95 and TNFR1 contains the 'death domain' (light grey box) which signals by homotypically interacting with FADD/MORT1. The intracellular region of TNFR2 contains the 'TRAF' domain (stippled dark grey box) which signals by interacting with various TNFR-associated factors (TRAFS).

patients that have been studied at the National Institutes of Health in the USA, 17 unrelated families have at least one member who fits the clinical definition of ALPS.

In 13 of the ALPS families, we have detected mutations in the CD95 receptor in the proband and in several members of their families. In only one family is there evidence for a new mutational event in which the proband harbours a CD95 mutation which is lacking in both biological parents. In all other families, the CD95 mutations are inherited. These mutations either lead to premature termination of the protein-coding sequence or they cause amino acid substitutions (a representative group is illustrated in Fig. 4) (Sneller et al 1997). The majority of CD95 mutations are found in the region of the gene that is known as the 'death domain' by virtue of the fact that excision of this part of the protein abrogates death signalling (Nagata & Golstein 1995, Cleveland & Ihle 1995). The death domain is also the docking site for the cytosolic adaptor protein FADD/MORT1 (Chinnaiyan & Dixit 1997) (see below). However, in each of our families with CD95 mutations, the mutations are heterozygous with a normal allele. This raises several questions: (1) why is apoptosis defective if wild-type CD95 receptor chains are present in the cells of heterozygous individuals?; (2) how do mutations in different locations throughout the death domain cause defective apoptosis?; and (3) why are there several clinically normal parents that transmit Fas mutations? The ability of heterozygous mutant CD95 chains to cause defective apoptosis appears to be due to the fact that CD95 and all members of the TNF receptor superfamily form homotrimeric receptors on the cell surface (Smith et al 1994). This appears to be essential for engaging the ligand which is also homotrimeric with an axis of symmetry that is the same as the receptor. We have directly shown that the transmission of death signals by wild-type CD95 molecules is potently inhibited by the presence of mutant CD95 molecules, suggesting that trimers that have as few as one defective CD95 molecule may be unable to signal (Fisher et al 1995). Among the cases of ALPS associated with CD95 mutations that have been reported world-wide, all except one have heterozygous mutant alleles (Fisher et al 1995, Rieux-Lancat et al 1995, Drappa et al 1996). In the homozygous case resulting from a consanguinous marriage, the mutation was a 'loss of function' allele in that it caused destabilization of the CD95 mRNA and no protein was produced (Rieux-Lancat et al 1995).

To understand how mutations dispersed throughout the death domain abrogate CD95 function, we have carried out structure–function analyses of various mutant CD95 proteins in collaboration with Drs Stephen Fesik and Baohua Huang at Abbott Research Laboratories. Based on their report of the death domain structure comprising a nest of six α helices (Huang et al 1996), we determined the structures of death domains containing the amino acid substitutions encoded by the ALPS mutations. The location and chemical nature of the amino acid changes

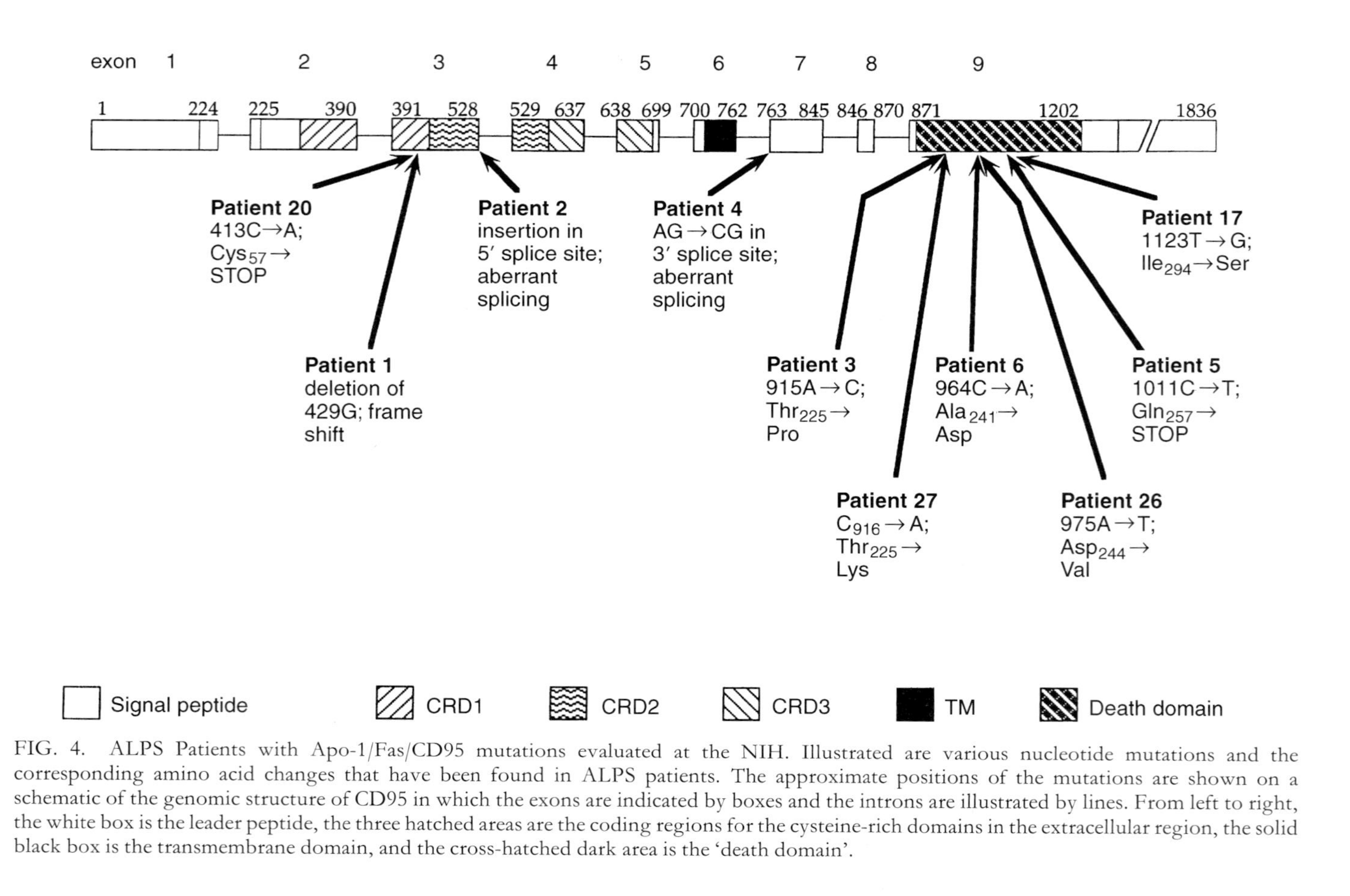

FIG. 4. ALPS Patients with Apo-1/Fas/CD95 mutations evaluated at the NIH. Illustrated are various nucleotide mutations and the corresponding amino acid changes that have been found in ALPS patients. The approximate positions of the mutations are shown on a schematic of the genomic structure of CD95 in which the exons are indicated by boxes and the introns are illustrated by lines. From left to right, the white box is the leader peptide, the three hatched areas are the coding regions for the cysteine-rich domains in the extracellular region, the solid black box is the transmembrane domain, and the cross-hatched dark area is the 'death domain'.

in ALPS caused them to have one of two possible structural effects. For mutations in α helices 4, 5, 6, in which hydrophobic residues are often replaced by more polar amino acids, there is a global unfolding of the death domain structure (patients 3 and 17). By contrast, several mutations that fall within the α-3 helix do not disturb the overall structure of the death domain (patients 26 and 6). A crucial role of the CD95 death domain is to bind a signal adaptor protein FADD (Fas-associated death domain protein) (Fig. 5). Each of the amino acid changes that were studied prevented FADD–CD95 death domain interactions. This allows us to conclude that the α-3 helix is likely to directly contact FADD such that local structural perturbations disrupt FADD binding. On the other hand, alterations in hydrophobic residues in other helices cause drastic changes in the overall death domain structure and therefore may directly or indirectly interfere with the FADD binding site. Thus both global and local changes in the Fas death domain structure prevent FADD interactions and lead to apoptosis defects. These results raise interesting and unresolved issues regarding the stoichiometry of FADD binding to the CD95 homotrimer and suggest the possibility that occupancy of at least two and possibly all three FADD interaction sites in the trimer could be necessary for death signal transmission. For example, the expectation is that only 1/8 homotrimers would contain exclusively mutant subunits (when combining an equivalent number of wild-type and mutant CD95 chains) which would seem to be insufficient to explain the substantial impairment of apoptosis. Further experiments are in progress to determine precisely how the mutant CD95 chains dominantly interfere with signal integration.

The transmission of deleterious CD95 alleles in a clinically silent manner has been an enigma. In certain ALPS cases, the parent, who carries the mutation but is clinically normal, can be demonstrated to have defective lymphocyte apoptosis when tested *in vitro* (see the results for the mothers of patients 2 and 4 in Fisher 1995). In many of these individuals, even scrutiny of their early clinical history does not turn up evidence of lymphoproliferation or autoimmunity. Testing is underway to determine whether there is a particular HLA associated with manifest disease. This is likely to be the case, as HLA alleles have been prominently associated with insulin-dependent diabetes, multiple sclerosis and other autoimmune diseases (Todd 1995). Moreover, there are likely to be other 'modifier' gene alleles that determine the clinical 'penetrance' of the CD95 defect. In mice, the *lpr* mutation has different clinical manifestations depending on the inbred genetic background on which it resides (Cohen & Eisenberg 1991).

Another issue of considerable interest is the fact that four of the ALPS patients have normal CD95 genes. This has prompted us to investigate other components of the CD95 signalling pathway that culminates in apoptosis, particularly caspase proteases. Caspases form a family of cysteine endoproteinases that require aspartate at the P1 site in the substrate protein (Miller 1997) (Fig. 5). Genetic investigation of

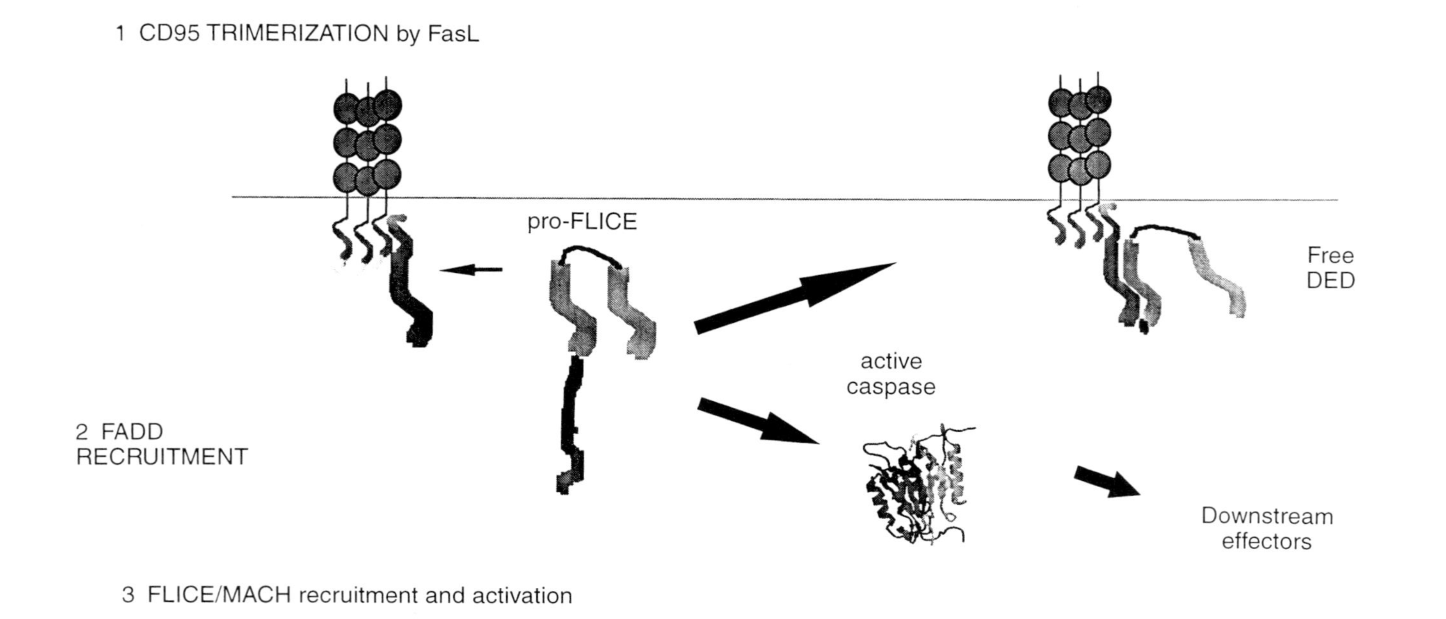

FIG. 5. Schematic of the CD95–FADD–FLICE signalling model. In this model CD95 trimerization leads to the recruitment of FADD followed by the binding of pro-FLICE (also referred to as MACH or caspase 10, see Miller 1997). This causes a processing event that results in the release of active caspase from the complex and the induction of downstream effector events that cause cell death.

the ALPS patients with normal CD95 genes has revealed that there also are no mutations in the CD95 ligand, FADD/MORT1 and FLICE/MACH1 (caspase 8) genes (J. Wang & M. J. Lenardo, unpublished results). However, in one such family, we have detected a mutation in a caspase gene. This mutation appears to alter the activity of this protease and render it incapable of inducing apoptosis. Preliminary results suggest that the mutant enzyme subunits can dominantly interfere with Fas-induced apoptosis. In the three remaining CD95 normal ALPS families, mutations in apoptotic molecules have yet to be found.

Summary

We have used molecular genetic approaches to understand immunological tolerance that is due to apoptosis of mature peripheral lymphocytes. We found that genetic deficiencies in apoptotic pathways are an important component of ALPS which comprises abnormalities in lymphocyte homeostasis and autoimmunity. These mutations are transmitted through families with extremely variable penetrance implying that other genetic and/or environmental influences play a prominent role in the disease. Further studies will be directed at elucidating other genes that are likely to play a major role in the pathogenesis of ALPS as well as other, more prevalent, autoimmune diseases.

References

Boehme SA, Lenardo MJ 1993 Propriocidal apoptosis of mature T lymphocytes occurs at S phase of the cell cycle. Eur J Immunol 23:1552–1560

Chinnaiyan AM, Dixit VM 1997 Portrait of an executioner: the molecular mechanism of Fas/APO-1-induced apoptosis. Immunology 9:69–76

Cleveland JL, Ihle JN 1995 Contenders in FasL/TNF death signalling. Cell 81:479–482

Cohen PL, Eisenberg RA 1991 *Lpr* and *gld*: single gene models of systemic autoimmunity and lymphoproliferative disease. Annu Rev Immunol 9:243–265

Critchfield JM, Boehme S, Lenardo MJ 1995 The regulation of antigen-induced apoptosis in mature T lymphocytes. In: Gregory CT (ed) Apoptosis and the immune response. Wiley-Liss Inc, New York, p 55–114

Drappa J, Vaishnaw AK, Sullivan KE, Chu J-L, Elkon KB 1996 *Fas* gene mutations in the Canale-Smith syndrome, an inherited lymphoproliferative disorder associated with autoimmunity. N Engl J Med 335:1643–1649

Fisher GH, Rosenberg FJ, Straus SE et al 1995 Dominant interfering Fas gene mutations impair apoptosis in a human autoimmune lymphoproliferative syndrome. Cell 81:935–946

Huang B, Eberstadt M, Olejniczak ET, Meadows RP, Fesik SW 1996 NMR structure and mutagenesis of the Fas (APO-1/CD95) death domain. Nature 384:638–641

Kang S-M, Beverly B, Tran A-C, Brorson K, Schwartz RH, Lenardo MJ 1992 Transactivation by AP-1 is a molecular target of T cell clonal anergy. Science 257:1134–1138

Lenardo MJ 1991 Interleukin-2 programs mouse $\alpha\beta$ T lymphocytes for apoptosis. Nature 353:858–861

Lenardo MJ 1996 Fas and the art of lymphocyte maintenance. J Exp Med 183:721–724
Miller DK 1997 The role of the caspase family of cysteine proteases in apoptosis. Immunology 9:35–49
Nagata S 1994 Mutations in the Fas antigen gene in *lpr* mice. Semin Immunol 6:3–15
Nagata S, Golstein P 1995 The Fas death factor. Science 267:1449–1456
Rieux-Laucat F, LeDeist F, Hivroz C et al 1995 Mutations in Fas associated with human lymphoproliferative syndrome and autoimmunity. Science 268:1347–1349
Smith CA, Farrah T, Goodwin RG 1994 The TNF receptor superfamily of cellular and viral proteins: activation, costimulation, and death. Cell 76:959–966
Sneller MC, Straus SE, Jaffe ES et al 1992 A novel lymphoproliferative/autoimmune syndrome resembling murine *lpr/gld* disease. J Clin Invest 90:334–341
Sneller MC, Wang J, Dale JK et al 1997 Clinical, immunologic, and genetic features of an autoimmune lymphoproliferative syndrome associated with abnormal lymphocyte apoptosis. Blood 89:1341–1348
Todd JA 1995 Genetic analysis of type 1 diabetes using whole genome approaches. Proc Natl Acad Sci USA 92:8560–8565
Zheng L, Fisher G, Miller RE, Peschon J, Lynch DH, Lenardo MJ 1995 Induction of apoptosis in mature T cells by tumour necrosis factor. Nature 377:348–351

DISCUSSION

Kioussis: I have a genetic question about ALPS. If a mutant CD95 allele causes a disease in a heterozygous state, what pressure maintains it in the population? Does such an allele represent only a spontaneous mutation which is maintained for one or two generations and then disappears, or can you trace it back in the family tree?

Lenardo: Only one of the 17 ALPS patients appears to be the result of a new mutation. In all the other cases, we can track the mutation through at least two or three generations. In two of the new families that we have studied, the disease can be traced through five generations and we are attempting to positively identify the CD95 mutation in as many of these individuals as we can, given the specimens that we obtain. In deceased individuals it is difficult because even tissue blocks are not available. The mutation does not affect fertility. Usually, in adulthood the disease actually gets better, probably because of thymic involution and there's not as much pressure from new T cell production in the system—it is similar to what happens if you thymectomize an *lpr* mouse: they tend to get better. But in some cases, in bad alleles that tend to be fully penetrant, they can be associated with malignancy.

Cornall: Do you have any idea what the gene frequency might be?

Lenardo: Not exactly, but ALPS patients are not hard to find: we have a new family coming in roughly every two months. I know there's a group in Italy that's following several ALPS-like families, although they have not yet identified CD95 mutations in their children (Dianzani et al 1997), a group in France that has several ALPS families (Le Deist et al 1996), and a group in New York that has studied a couple of ALPS families (Drappa et al 1996).

Cornall: Are there mutations shared between families that might have originated from a single mutation event?

Lenardo: Probably, because in only one case out of 13 that we've been able to document is it a novel mutation. Since the mutation doesn't affect fertility, people with CD95 mutations are procreating and spreading the alleles.

Cornall: By linkage disequilibrium you could see whether there's an association with a linked marker. You might be able to extrapolate backwards from this.

Lenardo: We're starting pretty much in the same place that you guys started: we're looking at HLA. The early results are promising but they have to be replicated in independent families.

Cornall: Do you think the Fas-related mutations are wholly deleterious, if they are indeed so common?

Lenardo: I would say they're not highly deleterious, unless you get one of the bad alleles that leads to a lymphoma.

Cornall: Do you think the alleles could be positively selected?

Lenardo: We don't have any evidence for that.

Mason: Do these individuals make better responses to foreign antigens? If you wanted to look for a selective advantage, then a better than average immune response might be it.

Lenardo: They make better responses against their own platelets and red cells! We haven't looked further than that.

Allen: They must get childhood vaccines.

Lenardo: That's a good point: it would be interesting to look at that.

Abbas: Pre-disease *lpr* mice make normal responses to foreign protein antigens.

Jenkins: Along these lines, how do you think that pathway fits into the normal response? Does it have something to do with this big loss in effector cells that occurs after the peak in clonal expansion?

Lenardo: I think Abul put it the same way that I think about it, which is that if you have some arbitrary number of T cells you start out with, when stimulation occurs at time zero, the observation in many systems is that the T cell number increases and, at some later time, returns to baseline. I think the mechanism that primarily accounts for this cell loss is what Abul called 'passive death' and what I call 'lymphokine withdrawal apoptosis'. In fact, a beautiful experiment was done on this phenomenon with staphylococcal enterotoxin B (SEB) by a group in Japan (Kuroda et al 1996). Whereas we gave repeated SEB injections every two days and that's when you can really demonstrate the active or propriocidal death pathway because you're driving the system (cf. Lenardo 1991), if you give a single dose, as Kuroda and colleagues did, the T cell number goes up. Then, over the course of a couple of weeks, it decreases, although sometimes it actually goes a little bit longer (Lenardo 1991). What Kuroda and colleagues did was to implant a continuous infusion pump that was releasing IL-2 at the peak of the proliferative response, and they showed that the cell number did not decrease. That to me is the most convincing experimental evidence that the decrease after a single administration of SEB is due to cytokine withdrawal. Therefore both the propriocidal and the passive or lymphokine withdrawal pathways can be modelled nicely *in vitro* and they are molecularly very distinct.

Kurts: The patients you described had mutations in Fas. Are there also patients with a mutation in Fas ligand, similar to the situation in *gld* mice? And if so, would male patients have fertility problems? For mice it was reported that the immunoprivileged situation of the testis is based on its expression of Fas ligand (Bellgrau et al 1995).

Lenardo: That's an interesting question. One patient identified among a group of systemic lupus erythematosus patients by John Mountz in Alabama, has a Fas ligand mutation (Wu et al 1996). This is the only one we know of in the world. Clearly these mutations are less frequent, on the basis of the existing data.

Hämmerling: Were they homozygous for this mutation?

Lenardo: I don't think so.

Hämmerling: So dose effect is sufficient to produce a phenotype.

Abbas: That's true for the *gld* mouse, too. Heterozygotes are abnormal.

Hämmerling: How abnormal are they?

Abbas: They don't get frank autoimmune disease, but you can find defects in apoptosis.

Lenardo: I think it depends on the genetic background. In the patient described by Mountz and co-workers (Wu et al 1996), we thought that the clinical history sounded like ALPS. Although this individual had made it into a group of patients who were considered lupus patients, they had adenopathy, which is unusual in lupus. We have never really had access to that patient and their material, so we haven't investigated them.

Simon: Do you think that viruses that can produce natural dominant negative forms of caspase 8/FLICE or other molecules recruited on the death pathway, could have something to do with other autoimmune diseases?

Lenardo: Clearly it is to the benefit of the virus to prevent apoptosis so that death of the cell is postponed until new virions are produced. In collaboration with John Bertin and Jeff Cohen, we characterized viral inhibitors that were identified on the basis of the fact that they have domains that are homologous to the 'death effector domain' which is found in FADD, caspase 8 and caspase 10. This domain allows the viral proteins to competitively bind these proteins and prevent the formation of the signalling complex that is crucial for CD95 and TNF apoptosis.

Mitchison: What about the heterozygous Fas mutants in the mouse?

Abbas: They are normal, because it's a production defect not a structural defect.

Lenardo: There's no RNA and there's no protein from the *lpr* allele, and so as a consequence, if you have one good allele you can get some receptor on the surface that is completely wild-type.

Waldmann: With regard to those mutants in the hydrophobic region, if you look on the cell surface to see how much Fas is expressed, what do you see?

Lenardo: Surprisingly, it looks pretty good.

Waldmann: You would think from a Poisson distribution that you would get quite a few competent molecules on the cell surface, so why with the heterozygotes do you see a phenotypic defect?

Lenardo: I think it would be one-half to the third power that would be expected based on probability alone that would contain exclusively wild-type subunits, and that doesn't seem to be enough.

Mitchison: Do you think it is as simple as that?

Lenardo: I hesitate to say more, because biochemically it is difficult to prove which is the point mutant and which is not. There are ways you can do this, by tagging with green fluorescent protein, for instance, but then you have altered the protein. I haven't been enthusiastic about doing that, because then it's not the real Fas mutant.

Mitchison: It is obviously too early to generalize, but it is tempting to predict that there will be dominant mutations in, for example, the IL-2 receptor, which is trimeric, but not in the PDGF receptor, which is dimeric. In the latter case you would then have a quarter of the molecules intact in the heterozygotes, and you might expect that to be enough. This form of dominance is a remarkable phenomenon.

Lenardo: We think about Fas in the way that Paul Allen thinks about the T cell receptor (TCR)—that is, to really get the signal through, you may need higher-order structures. So even if you have one or two subunits in the trimer that are wild-type, it may need to get into a larger structure and this might not occur if seven out of eight complexes on the surface of the cell do not allow this function. So there may be another level of complexity we haven't looked at.

Mitchison: You could almost turn the argument on its head and say if, for example, the TCR is assembled into large aggregates (which some people think is the case) then you should see lots of dominant mutations in TCR receptors. These haven't been seen, so perhaps those aggregates are a figment of the imagination.

Wraith: But cells bearing these receptors would never be positively selected for in the thymus.

Hämmerling: To come back to the beginning of your talk, do you know how the IL-2 phenomenon works? If you first give IL-2 and then come with the antigen, the T cells will die. So what does the IL-2 do? Does it induce the formation of these large structures which have just been discussed?

Lenardo: We haven't completely solved this. One thing that we're quite sure about is that a lot of this has to do with cell cycle progression. For example, there is the old phenomenon where hybridomas which are cycling by virtue of an oncogene are very susceptible to endogenous apoptosis by these pathways when you cross-link TCRs. If you block cells which are cycling in IL-2 or IL-7 early in G1 they're protected, so clearly they have to move beyond this point in G1 in order to become susceptible. It also turns out when cells are cycling in IL-2, there's a dramatic up-regulation in the level of mRNA for the death cytokines and their receptors, particularly the ligands, when you cross-link TCRs. So although they are induced in the naïve resting cell, by gene regulation standards it's not impressive: you get about a three- and fivefold induction. After IL-2 you get between an 80–300-fold induction. Beyond that we don't know what it is about the cell cycle that predisposes T cells to Fas or TNF-induced apoptosis. One thing we have tried is to put in dominant negative cyclin-dependent kinases to block the susceptibility. We have also shown that direct Fas cross-linking works a lot more efficiently if the cells are cycling beyond early G1. There's clearly a molecular connection there that we haven't uncovered yet. With respect to the up-regulation of mRNAs, we haven't looked at the mechanism behind that.

Waldmann: Are antibodies to CD2 and class I MHC, which also give activation-induced cell death, also working in the same way?

Lenardo: We and others have never been able to reproduce the class I MHC work from Richard Miller (Sambhara & Miller 1991). CD2 does kill in the same manner as antigen stimulation because it delivers something akin to a TCR signal and you get Fas ligand up-regulation.

References

Bellgrau D, Gold D, Selawry H, Moore J, Franzusoff A, Duke RC 1995 A role for CD95 ligand in preventing graft rejection. Nature 377:630–632

Dianzani U, Bragardo M, DiFranco D et al 1997 Deficiency of the Fas apoptosis pathway without *Fas* gene mutations in pediatric patients with autoimmunity/lymphoproliferation. Blood 89:2871–2879

Drappa J, Vaishnaw AK, Sullivan KE, Chu J-L, Elkon KB 1996 *Fas* gene mutations in the Canale-Smith syndrome, an inherited lymphoproliferative disorder associated with autoimmunity. N Engl J Med 335:1643–1649

Kuroda K, Yagi J, Imanishi K et al 1996 Implantation of IL-2-containing osmotic pump prolongs the survival of superantigen-reactive T cells expanded in mice injected with bacterial superantigen. J Immunol 157:1422–1431

Le Deist F, Emile JF, Rieux-Laucat F et al 1996 Clinical, immunological, and pathological consequences of Fas-deficient conditions. Lancet 348:719–723

Lenardo MJ 1991 Interleukin-2 programs mouse $\alpha\beta$ T lymphocytes for apoptosis. Nature 353:858–861

Sambhara SR, Miller RG 1991 Programmed cell death of T cells signaled by the T cell receptor and the α3 domain of class I MHC. Science 252:1424–1427
Wu J, Wilson J, He J, Xiang L, Schur PH, Mountz JD 1996 Fas ligand mutation in a patient with systemic lupus erythematosus and lymphoproliferative disease. J Clin Invest 98:1107–1113

General discussion II

Genetics of immunological disease

Mitchison: I would like to follow up Mike Lenardo's talk to deal with the issue that was raised of the genetics of these diseases, concerning the nature of the variation which is important in their control. To start with, it seems to me that these rare mutations, which are of enormous biochemical interest, probably make only a minor contribution to human variation in disease susceptibility. Take for example rheumatoid arthritis, where the current estimate is that 30% of the susceptibility is in the genetics. That 30% will be mainly genetics somewhere in the centre of the bell-shaped curve of variation, where it is probably very different from variation at the tail ends where these structural gene mutations are involved. It seems reasonable to speculate on the basis of what we know at present that most of that variation will be in regulatory gene segments controlling the 'introvert proteins' of the immune system as distinct from the 'extrovert' ones which see antigen and so on. For example, the genetics of IL-1, IL-1 receptor and IL-1 receptor antagonist (IL-1ra), which Gordon Duff and his colleagues in Sheffield have looked at in great detail, seems to be almost entirely variation in the regulatory gene segments (Mitchison 1997). At a recent Gordon Conference, Linda Wicker described an *idd* gene (for susceptibility to insulin-dependent diabetes) in the mouse which is a structural variant in IL-2, but that seems to be exceptional, and not evident in the human population. These are broad generalizations, but isn't that the way we feel that the genetics is going? Polymorphisms in regulatory gene segments seem likely to be responsible for most of the genetic variation in disease susceptibility.

Abbas: I think that's the way it's going, but we currently have very limited information.

Hafler: It is very problematic, because in one instance two genes may predispose to autoimmune disease, but a another pair of genes may counteract their influence.

Mitchison: Am I not right in thinking that if you take quantitative trait loci (QTLs) as a whole, hardly any have yet been shown to be associated with any structural gene?

Cornall: I would like to mention obese diabetic mice, which represent a similar challenge. So far, at least 10 genetic susceptibility loci have been found in genetic analyses of NOD mice, including the MHC. Most of these loci are yet to be mapped, and this is made difficult because the effects of different alleles are invariably penetrant and different combinations of alleles can contribute to similar disease risk. Interestingly, two of the diabetes susceptibility genes identified by John Todd in his original NOD × C57BL/10 backcross came from the resistant strain. The overall picture of susceptibility to polygenic autoimmune disease is one in which multiple genetic variants act as QTLs in a way that generates disease risk as a quantitative trait. These QTLs must be acting in the very sorts of pathways that we have been discussing at this meeting. The high frequency of the QTL variants in human populations must be evidence of their selection, presumably by immune advantage (Vyse & Todd 1996).

Hafler: Vijay Kuchroo did an experiment with Linda Wicker, using the Idd3 locus congeneic mouse. Idd3 is protective in diabetes and it also appears to be protective in experimental autoimmune encephalomyelitis (EAE) (V. K. Kuchroo, personal communication). These are preliminary experiments but it does appear to have a major effect on the induction of EAE.

Mitchison: That also comes clearly out of Elaine Remmers' study of rat collagen-induced arthritis (Remmers et al 1996), where a number of the genes were repeats of genes which were picked up previously in the diabetes or EAE studies. A large number of the QTLs are suspiciously close to either cytokines or cytokine receptors.

Allen: But aren't we forgetting the MHC? $H2^{g7}$ is a dominant gene. Therefore, thymic editing and the failure of central tolerance is still going to be the key underlying effect. These other ones are important, but isn't that still the main one?

Mitchison: That's true.

Allen: The issue still remains as to how to look at a T cell repertoire.

Cornall: The risk of insulin-dependent diabetes in sibs of affected individuals is 15 times greater than in sibs of unaffected individuals. This gives a measure of the familial clustering, which may be due to genetic or environmental effects. Only part of this familial clustering is due to susceptibility at the MHC (Vyse & Todd 1996).

Hafler: Similar effects are seen in all autoimmune diseases, with genetic susceptibility accounting for 30–40% risk.

Mitchison: The multiple sclerosis (MS) studies are at an interesting stage. There have been two large surveys, one British and another American, which agreed that the disease is multifactorial. But there was disagreement about all the other genes, apart from the MHC, which was identified in both (Sawcer et al 1996, Haines et al 1996).

Mason: Presumably infectious diseases keep these polymorphisms in the population. Are there any MHC associations with resistance to certain types of infectious disease?

Lechler: Malaria and B53: that is about it.

Mitchison: What about leprosy? I don't think there is any shortage of studies. In any case, Don, why do you want human data? Why can't you accept chickens and mice? Marek's disease in chickens is strongly MHC associated.

Perhaps a more interesting question concerns what maintains cytokine receptor and cytokine receptor antagonist polymorphisms in the population. I would suggest that this is likely to be basically the same mechanisms as operate in HLA: they provide flexibility and enable appropriate responses to be made against a wider range of infections than do homozygotes.

Mason: We should remember that too much of an immune response is also a bad thing. The reason people die of tuberculosis is not because of the disease organism but as a consequence of the body's response to it.

Mitchison: Absolutely. If it was just the absolute levels of cytokines which were being selected that wouldn't make very much sense. But if you have two alleles with two different promoters, cells can turn them on and do one thing in one situation and something else in another. Let me give you an example of that. Villard et al (1996) in Paris, have studied extensively the phenotypic effect of the strong regulatory mutants in the angiotensin I converting enzyme. Their suggestion, with a certain amount of evidence, is that whereas high levels are bad for blood pressure, they may be useful locally in wound healing.

Lenardo: One point that I think is critical on this issue is that when you talk about these loci, they're not disease-causing loci in the sense of a simple Mendelian inherited disorder: rather, they are susceptibility loci. Tying in with the point I think you're getting to, in an individual any one of these loci in isolation might promote the expression of a cytokine or some other molecule that's good for a particular kind of infection, so it stays in the population, but then it's the confluence of all loci together which is what is bad by virtue of the fact that they cause autoimmune disease. However, even if you have the full group of the susceptibility loci in any given individual, you may still not get the disease: they only create a susceptibility. In diabetes, there are well described susceptibility loci that even when they occur together still do not cause the disease—they merely create the susceptibility.

Hafler: But in identical twins with islet cell anti-insulin antibodies, the risk of developing diabetes is almost 100%.

Lenardo: But there are identical twins that have the same genetic complement in which one gets sick and the other does not. It is still a susceptibility effect and not determinitive for disease without other influences, perhaps arising in the environment.

Mitchison: Of course, the genetics of disease susceptibility is no different in this respect from any other genetic trait.

References

Haines JL, Ter MM, Bazyk A et al 1996 A complete genomic screen for multiple sclerosis underscores a role for the major histocompatibility complex. Nat Genet 13:469–471

Mitchison NA 1997 Partitioning of genetic variation between regulatory and coding gene segments: the predominance of software variation. Immunogenetics 46:46–52

Remmers EF, Longman RE, Du Y et al 1996 A genome scan localises five non-MHC loci controlling collagen-induced arthritis in rats. Nat Genet 14:82–85

Sawcer S, Jones HB, Feakes R et al 1996 A genome screen in multiple sclerosis reveals susceptibility loci on chromosome 6p21 and 19q22. Nat Genet 13:464–468

Villard E, Tiret L, Visvikis S, Rakotovao R, Cambien F, Soubrier F 1996 Identification of new polymorphisms of the angiotensin I converting enzyme (ACE) gene, and study of their relationship to plasma ACE levels by two-QTL segregation-linkage analysis. Am J Hum Genet 58:1268–1278

Vyse TJ, Todd JA 1996 Genetic analysis of autoimmune disease. Cell 85:311–318

A role for CTLA-4-mediated inhibitory signals in peripheral T cell tolerance?

James P. Allison, Cynthia Chambers, Arthur Hurwitz, Tim Sullivan, Brigitte Boitel, Sylvie Fournier, Monika Brunner and Matthew Krummel

Howard Hughes Medical Institute, Department of Molecular and Cell Biology and the Cancer Research Laboratory, University of California, Berkeley, CA 94720-3200, USA

Abstract. Occupancy of the antigen receptor is not sufficient for activation of naïve T cells—additional co-stimulatory signals are required that can be provided only by 'professional' antigen-presenting cells. This two-signal model for T cell activation has been thought to provide a mechanism for the induction and maintenance of peripheral tolerance. Work over the past six years has demonstrated that the relevant co-stimulatory receptor on T cells is the molecule CD28. Recent data shows that the CD28 homologue CTLA-4 plays a role in negative regulation of T cell responses. Here we suggest that CTLA-4 may also serve as an attenuator of T cell-activating signals, raising the threshold of stimulation required to obtain full activation. The inhibitory signals mediated by CTLA-4 may provide an additional mechanism for the maintenance of peripheral tolerance.

1998 Immunological tolerance. Wiley, Chichester (Novartis Foundation Symposium 215) p 92–102

The two-signal model of T cell activation stems from work carried out by Mark Jenkins and Ron Schwartz showing that antigen-presenting cells (APCs) express chemically labile co-stimulatory signals that are required in addition to antigen receptor signals to obtain activation of T cell clones. In the absence of these second signals, T cell clones were rendered anergic (Jenkins & Schwartz 1987). The demonstration that these co-stimulatory signals were restricted to professional APCs, that is, dendritic cells, macrophages and activated B cells, and were not expressed by epithelial or endothelial cells, led to the proposal that this two-signal model could provide a mechanism for understanding induction and maintenance of peripheral T cell tolerance (Jenkins et al 1988, Mueller et al 1989). Work from a number of labs established that the major receptor for co-stimulatory signals on the T cell is the molecule CD28, and that members of the B7 family are the APC counter-receptors on APCs (Jenkins & Johnson 1993, Linsley & Ledbetter 1993, Allison 1994). It has become generally accepted that a CD28 signal is required for activation of naïve T cells.

Recently, this scenario became complicated by the addition of another actor in the process: CTLA-4. The CTLA-4 gene was initially isolated in a subtractive screen for genes involved in the function of cytotoxic T cells, although CTLA-4 is expressed in both $CD4^+$ and $CD8^+$ cells after activation (Brunet et al 1987). CTLA-4 is a close homologue of CD28, has a very similar genomic organization, and is genetically linked to the CD28 gene (Harper et al 1991). Studies with soluble chimeric fusion proteins showed that CTLA-4, like CD28, bound to members of the B7 family but with affinities 20–50-fold higher (Linsley et al 1991, 1994). In the absence of antibodies with which to assess its distribution and function, the role of CTLA-4 in T cell activation has, until recently, remained obscure. The question is: why do T cells, which have a perfectly functional APC molecule in CD28 that is constitutively expressed at reasonably high levels, also express an additional B7 counter-receptor only after activation with a much higher affinity for those counter-receptors?

Recent generation of antibodies to CTLA-4 has shed some light on this issue. Initial experiments using an antibody to human CTLA-4 demonstrated that while the anti-CTLA-4 alone had no effect in co-stimulation assays, enhancement of T cell responses was observed when anti-CTLA-4 was added together with anti-CD28 (Linsley et al 1992). This led to the notion that the function of CTLA-4 was to synergize with CD28 after T cell activation and to sustain co-stimulation and allow continued production of interleukin IL(-2) and proliferation of T cells, even in the face of diminishing B7-mediated co-stimulation. However, work from other labs suggested that apparent enhancement of T cell responses in these assays might not be a result of synergistic co-stimulation, but is rather the result of blockade of inhibitory signals mediated by CTLA-4/B7 interactions in the cultures (Walunas et al 1994, Krummel & Allison 1995). Supporting this idea was the demonstration that co-ligation of CTLA-4 in combination with T cell receptor and CD28 resulted in an inhibition of T cell responses (Krummel & Allison 1995). It was also observed *in vivo* that while injection of anti-CD28 antibodies inhibited responses of T cells to the superantigen staphylococcal enterotoxin B, administration of antibodies to CTLA-4 enhanced those responses (Krummel et al 1996). Together these data were taken as supporting a down-regulatory role of CTLA-4 in T cell activation.

While these results were initially controversial, the issue seems to have been resolved by analysis of CTLA-4-deficient mice in three different laboratories (Tivol et al 1995, Waterhouse et al 1995, Chambers et al 1997). These mutant mice show a common phenotype: an extreme lymphadenopathy with marked T cell infiltration of the tissues and lethality by three weeks of age. The T cells from these animals proliferate spontaneously and produce high levels of cytokines. The extreme phenotype of these mice supports the notion that CTLA-4 plays an important role in limiting T cell activation.

The basis for the lymphadenopathy in the CTLA-4-deficient mice remains to be determined, but at least two non-exclusive mechanisms can be suggested (Allison & Krummel 1995). The first stems from the observation that CTLA-4 is expressed at high levels only after T cell activation (Linsley et al 1992, Walunas et al 1994, Krummel & Allison 1995). According to this model, CTLA-4 is important in terminating T cell responses, and the phenotype of the deficient mice is a result of a failure to terminate ongoing responses to environmental antigens. The second possibility is that CTLA-4 is involved in regulating initiation of responses, and that the lymphoproliferation in the deficient mice is a result of inappropriate initiation of responses. This possibility is supported by several observations. While CTLA-4 is not detectable on the cell surface until 48 h or so after T cell activation, its effects in co-ligation assays can be detected much earlier (Krummel & Allison 1996). Inhibition of expression of CD69 can be detected as early as 12 h following activation, accumulation of interleukin (IL)-2 in culture supernatants can be detected as early as 16 h, and inhibition of CD25 can be detected as early as 24 h. Indeed, the kinetics of the appearance of IL-2 in cultures suggest that the inhibition of its production is not due to early termination, but rather to a failure of initiation. More recent data have shown that cross-linking of CTLA-4 can indeed prevent initiation of transcription of the IL-2 gene and is detectable as early as 4 h after activation (Brunner & Allison 1998). These results indicate that CTLA-4 may be operative at very early stages in T cell activation.

Blockade experiments provide direct evidence that CTLA-4 can regulate the outcome of T cell activation under some circumstances. We were interested in determining the basis for the inability of small resting B cells to activate efficiently naïve T cells despite the fact that unactivated B cells express low, but detectable, levels of B7–2. The question we sought to address was whether this low level was not sufficient to support CD28-mediated co-stimulation, or whether CTLA-4 might be inhibiting the response (Chambers et al 1996). Small resting B cells were pulsed with the appropriate peptide, and then used to stimulate naïve T cells from transgenic mice expressing the relevant T cell antigen receptor. We found that little activation was observed unless the cultures included non-stimulatory monovalent anti-CTLA-4 Fab fragments, whereupon we obtained vigorous proliferation. This proliferation was blocked if Fab fragments of anti-CD28 antibodies were also included. This experiment was interpreted as an indication that small resting B cells would express sufficient levels of B7 for at least moderate co-stimulation of T cells were it not for the inhibitory effects of CTLA-4. The fact that anti-CD28 Fab blocked the augmentation of the responses by anti-CTLA-4 Fab provides evidence that the CTLA-4 reagents were not providing agonistic APC signals, but rather were functional in inhibiting CD28-driven co-stimulation.

This observation led us to propose the model for CTLA-4 function shown in Fig. 1. According to this model, a T cell that receives no APC signals via B7 will not be activated. A T cell receiving antigen receptor signals from a cell expressing low levels of B7 will also not be activated because CTLA-4, as a result of its higher affinity, can compete with CD28 for B7 and generate inhibitory signals which prevent activation. However, when a T cell encounters antigen on an activated APC which expresses large amounts of B7, the low levels of CTLA-4 become limiting and there is B7 available to bind CD28. Thus, co-stimulation and activation proceed. As a consequence of activation, those cells up-regulate CTLA-4 expression to high levels which are sufficient to provide dominant inhibitory signals that terminate the T cell response. The mechanism of generation of inhibitory signals by CTLA-4 is a matter under investigation and may involve recruitment of protein tyrosine phosphatases to the region of the cell involved in T cell receptor (TCR)-mediated generation of protein tyrosine kinase cascades (Marengere et al 1996).

This model suggests that CTLA-4 might play a role in attenuating signals involved in T cell activation, thereby raising the threshold of stimulation required for full T cell response. An extreme form of this proposal would suggest that one of the functions of CTLA-4 might be to provide protection against the inherent affinity of all TCRs in the body for self MHC molecules that arises as a consequence of positive selection. It has been shown that signals from MHC molecules are required for survival (Tanchot et al 1997). Thus, naïve T cells circulating in the body in the absence of antigen are not doing so in a vacuum, but seem to be receiving a constant level of stimulation by self MHC. It can be suggested that one of the roles of CTLA-4 is to prevent signals required for survival from resulting in activation and consequent autoimmunity. Another non-exclusive possibility is that the function of CTLA-4 is to take the noise out of the system and raise the threshold of signals required for full T cell activation beyond that which is present under normal conditions, thereby contributing to peripheral tolerance.

Several lines of experimental evidence suggest a role for CTLA-4 in autoimmunity. The most extensive experimental work has been in a mouse model for multiple sclerosis, experimental autoimmune encephalomyelitis (EAE), that can be induced in susceptible strains of mice (PL, SJL, etc.) by immunization with peptides of proteolipid protein (PLP) in complete Freund's adjuvant (McRae et al 1992). Disease pathogenesis is characterized by induction of a Th1 T cell response and infiltration of the CNS which presents clinically as a paralytic syndrome that usually resolves itself. Our laboratory and others have reported that blockade of CTLA-4 with antibodies during immunization with PLP greatly exacerbates the disease both in terms of CNS infiltration and clinical symptoms (Karandakar et al 1996, Perrin et al 1996, Hurwitz et al 1997). In fact, we have demonstrated that

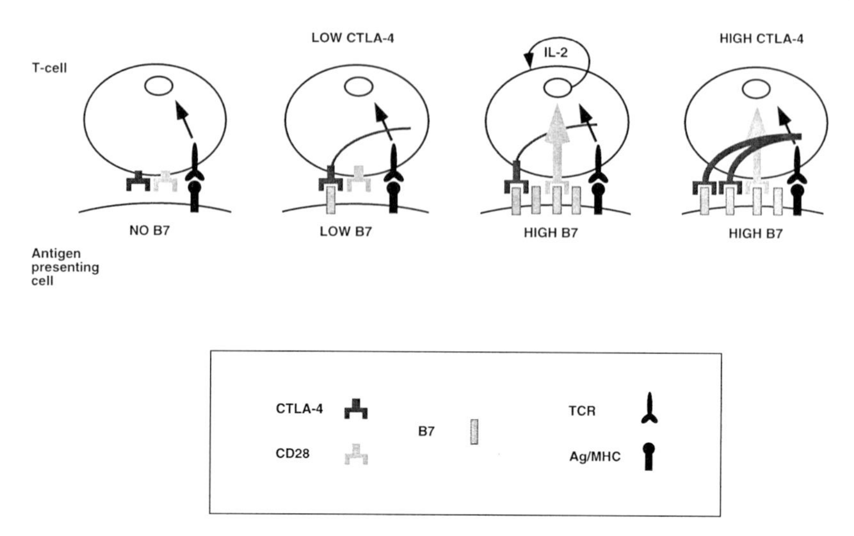

FIG. 1. Model of integration of TCR-, CD28- and CTLA-4-mediated signals. Under conditions of low B7 concentration, the high affinity of CTLA-4 results in the dominance of inhibitory signals. Under conditions of abundant B7, CTLA-4 becomes limiting and CD28-mediated co-stimulation dominates, resulting in full T cell activation. As a consequence of activation, CTLA-4 expression is increased, and inhibition again dominates, thus terminating the T cell response.

disease severity is enhanced from moderate paralysis (in control animals) to morbidity in mice treated with anti-CTLA-4. Recent data suggest an even more pivotal role for CTLA-4. We have found that CTLA-4 blockade promotes induction of EAE even in strains of mice that are normally resistant. For example, immunization of BALB/c mice with spinal cord homogenate normally does not result in induction of CNS infiltration or clinical disease. However, when CTLA-4/B7 interactions are blocked, BALB/c mice develop marked imflammatory foci in the CNS as well as the accompanying paralytic disease. One interpretation of this result is that blockade of CTLA-4 has allowed the breaking of peripheral tolerance in the mice, resulting in disease induction. Alternatively, CTLA-4 blockade may render encephalitogenic T cells more auto-aggressive, capable of crossing the blood–brain barrier and inducing demyelination.

Another model in which the role of CTLA-4 has been examined is a TCR transgenic model of diabetes (Luhder et al 1998). In this model, young mice do not normally develop diabetes. However, mice injected with anti-CTLA-4 months before they would develop diabetes universally developed diabetes that was marked by an aggressive T cell infiltrate of the pancreatic islets. Interestingly, these effects were observed only during a narrow time window before the onset of insulitis.

Summary

Recent data indicating an inhibitory role for CTLA-4 in the regulation of T cell activation have complicated the simple two-signal model of T cell activation and its application to the maintenance of peripheral tolerance. The T cell seems to integrate the stimulatory signals generated by the antigen receptor, APC signals mediated by CD28, and inhibitory signals transduced by CTLA-4. The relative levels of the signals is a reflection of the affinity and the amount of the antigen/ MHC complex, the level of B7 expressed by the APC, and the amount of CTLA-4 in the T cell. The latter two factors are influenced by the activation status of the APC and the T cell, respectively. It is our suggestion that as a result of this complex signal integration, CTLA-4 provides an additional mechanism to regulate initiation of T cell responses under conditions of suboptimal signalling, and thus plays a role in the maintenance of peripheral tolerance.

Acknowledgements

This work was supported by NIH grant #CA40041 from the National Cancer Institute and the Keck Foundation.

References

Allison JP 1994 CD28–B7 interactions in T-cell activation. Curr Opin Immunol 6:414–419

Allison JP, Krummel MF 1995 The Yin and Yang of T cell costimulation. Science 270:932–933

Brunet JF, Denizot F, Luciani MF et al 1987 A new member of the immunoglobulin superfamily CTLA-4. Nature 328:267–270

Brunner M, Allison JP 1998 CTLA-4-mediated inhibition of T cell cycle progression action at two levels. In preparation

Chambers CA, Krummel MF, Boitel B et al 1996 The role of CTLA-4 in the regulation and initiation of T cell responses. Immunol Rev 153:27–46

Chambers CA, Cado D, Truong T, Allison JP 1997 Thymocyte differentiation occurs normally in the absence of CTLA-4. Proc Natl Acad Sci USA 94:9296–9301

Harper K, Balzano C, Rouvier E, Mattei MG, Luciani MF, Golstein P 1991 CTLA-4 and CD28 activated lymphocyte molecules are closely related in both mouse and human as to sequence, message expression, gene structure, and chromosomal location. J Immunol 147:1037–1044

Hurwitz AA, Sullivan TJ, Krummel MF, Sobel RA, Allison JP 1997 Specific blockade of CTLA-4/B7 interactions results in exacerbated clinical and histologic disease in an actively-induced model of experimental allergic encephalomyelitis. J Neuroimmunol 73:57–62

Jenkins MK, Johnson JG 1993 Molecules involved in T-cell costimulation. Curr Opin Immunol 5:361–367

Jenkins MK, Schwartz RH 1987 Antigen presentation by chemically modified splenocytes induces antigen-specific T cell unresponsiveness *in vitro* and *in vivo*. J Exp Med 165:302–319

Jenkins MK, Ashwell JD, Schwartz RH 1988 Allogeneic non-T spleen cells restore the responsiveness of normal T cell clones stimulated with antigen and chemically modified antigen-presenting cells. J Immunol 140:3324–3330

Karandakar NJ, Vanderlugt CL, Walunas TL, Miller SD, Bluestone JA 1996 CTLA-4: a negative regulator of autoimmune disease. J Exp Med 184:783–788

Krummel MF, Allison JP 1995 CD28 and CTLA-4 deliver opposing signals which regulate the response of T cells to stimulation. J Exp Med 182:459–465

Krummel MF, Allison JP 1996 CTLA-4 engagement inhibits IL-2 accumulation and cell cycle progression upon activation of resting T cells. J Exp Med 183:2533–2540

Krummel MF, Sullivan TJ, Allison JP 1996 Superantigen responses and costimulation: CD28 and CTLA-4 have opposing effects on T cell expansion *in vitro* and *in vivo*. Int Immunol 8:519–523

Linsley PS, Ledbetter JA 1993 The role of the CD28 receptor during T cell responses to antigen. Annu Rev Immunol 11:191–212

Linsley PS, Brady W, Urnes M, Grosmaire LS, Damle NK, Ledbetter JA 1991 CTLA-4 is a second receptor for the B cell activation antigen B7. J Exp Med 174:561–569

Linsley PS, Greene JL, Tan P et al 1992 Coexpression and functional cooperativity of CTLA-4 and CD28 on activated T lymphocytes. J Exp Med 176:1595–1604

Linsley PS, Greene JL, Brady W, Bajorath J, Ledbetter JA, Peach R 1994 Human B7–1 (CD80) and B7–2 (CD86) bind with similar avidities but distinct kinetics to CD28 and CTLA-4 receptors. Immunity 1:793–801

Luhder F, Hoglund P, Allison JP, Benoist C, Mathis D 1998 Cytotoxic T lymphocyte-associated antigen 4 (CTLA-4) regulates the unfolding of autoimmune diabetes. J Exp Med 187:427–432

Marengere LEM, Waterhouse P, Duncan GS, Mittrucker HW, Feng GS, Mak TW 1996 Regulation of T cell receptor signaling by tyrosine phosphatase SYP association with CTLA-4. Science 272:1170–1173

McRae BL, Kennedy MK, Tan LJ, Dal Canto MC, Miller SD 1992 Induction of active and adoptive chronic-relapsing experimental autoimmune encephalomyelitis (EAE) using an encephalogenic epitope of proteolipid protein. J Neuroimmunol 38:229–240

Mueller DL, Jenkins MK, Schwartz RH 1989 Clonal expansion versus functional clonal inactivation: a costimulatory signalling pathway determines the outcome of T cell antigen receptor occupancy. Annu Rev Immunol 7:445–480

Perrin PJ, Maldonado JH, Davis TA, June CH, Racke MK 1996 CTLA-4 blockade enhances clinical disease and cytokine production during experimental allergic encephalomyelitis. J Immunol 157:1333–1336

Tanchot C, Lemonnier FA, Perarnau B, Freitas AA, Rocha B 1997 Differential requirements for survival and proliferation of CD8 naive or memory T cells. Science 276:2057–2062

Tivol EA, Borriello F, Schweitzer AN, Lynch WP, Bluestone JA, Sharpe AH 1995 Loss of CTLA-4 leads to massive lymphoproliferation and fatal multiorgan tissue destruction, revealing a critical negative regulatory role of CTLA-4. Immunity 3:541–547

Walunas TL, Lenschow DJ, Bakker CY et al 1994 CTLA-4 can function as a negative regulator of T cell activation. Immunity 1:405–413

Waterhouse P, Penninger JM, Timms E et al 1995 Lymphoproliferative disorders with early lethality in mice deficient in CTLA-4. Science 270:985–988

DISCUSSION

Jenkins: From what you have said, one would predict that it would be possible to grow out autoreactive T cells from the CTLA-4 knockout mouse. Has anybody been able to do that?

Abbas: We've tried to do that with Arlene Sharpe's CTLA-4 knockout. We have found that it is rather more difficult to grow out autoreactive T cells from the

CTLA-4 knockout than it is from *lpr* and *gld* mice. In the CTLA-4 knockout mice, the problem is determining what fraction of the cells is truly autoreactive. The only evidence that there is some true autoreactivity is that one can adoptively transfer those cells into Rag knockout mice and see the beginnings of infiltrates and myositis, so it's a transferable lesion (A. Sharpe, personal communication). This still doesn't formally address the question of autoreactivity — it's a tough problem.

Allison: I would argue that to the extent that they are autoreactive, it might not be because they are reacting to any tissue-specific antigens, but are responding to self MHC generally because the attenuation for activation provided by CTLA-4 is not available.

Shevach: What happens if you make hybridomas from the mouse right away and then look for reactivity against autologous antigens?

Abbas: We haven't done that yet, although for other reasons we are doing that with *lpr* and *gld* mice. But remember, even in these mice, if you take cells fresh out of the mouse and fuse right away, you don't see much autoreactivity: you have to run them through a cycle of *in vitro* stimulation with autologous APCs. You basically have to do an autologous mixed lymphocyte reaction and expand up this population.

Hämmerling: I have a question concerning the trafficking of the CTLA-4 molecule. Does AP50 bind to it in the ER?

Allison: No. Peter Linsley has actually shown that there is this rolling over of CTLA-4. You can't really see it on the cell surface, but if you add a fluorescence-tagged antibody to a T cell and incubate at 37 °C, the antibody will accumulate.

Hämmerling: What do you mean by 'rolling over'?

Allison: The protein comes to the ER and probably goes to the cell surface by normal routes, and is then taken into endosomes by the clathrin-coated pit machinery.

Hämmerling: Does it have a sorting signal which takes it to the cell surface and then internalizes again?

Allison: I don't think it gets internalized again.

Hämmerling: It seems to be expressed inside the endosome: but how does it get from the endosome to the T cell receptor?

Allison: I don't know, but it won't there while it is bound to AP50.

Wraith: A point of clarification relating to Gunter's question. Right at the beginning you said that upon activation, CTLA-4 was mobilized to the site of TCR engagement. Is that CTLA-4 on the cell surface or in the vesicle?

Allison: It seems to be in the vesicle. This is again work by Peter Linsley and Craig Thompson. They activated the T cells first so they could see it. They see vesicles which orient within minutes towards the side of the cell where the TCR is being engaged. You can see them actually traffic and the CTLA-4 then appears in the plaque.

Wraith: Is it fair to say that there is not enough on the cell surface to say whether or not it is mobilized at the cell surface?

Allison: That could be true.

Shevach: What happens to CTLA-4 in the CD28 knockout?

Allison: Levels at peak induction are lower, but it is still functional.

Shevach: But is it induced the same way? What I'm really asking is whether you need TCR engagement to get this to the cell surface, or is it a stochastic process?

Allison: You need TCR engagement to get CTLA-4 induced at sufficient levels to actually observe mRNA or protein. You can get that in the CD28 knockout mice although the levels are much reduced compared with intact mice.

Shevach: Why should CTLA-4 be reduced in the CD28 knockout?

Allison: Because there's a biphasic induction. If you look at mRNA, you get some very fast, but after about 24 h there is a second wave of induction that results in the much higher level that dominates the response later on.

Shevach: Is there an intracellular reservoir of CTLA-4 protein?

Allison: The real problem is that you can't detect the protein on unstimulated cells. There is no reservoir: even looking inside the cell you can't see it.

The point I want to make is that not being able to see it with the functional experiments says that there is enough there. You can imagine that there could be a timing lapse. If it is immobilized and if it recruits phosphatases it wouldn't take many molecules to muck up the whole thing. It is not a question of having to have hundreds of molecules on the cell surface so that you encounter B7 globally: rather, it's a question of firing it exactly where it's needed. However, the fact that we can't see it is annoying and make reviewers very uncomfortable.

Hafler: Do you think that in the endosomes the CTLA-4 binds B7–1 or B7–2 and thus compete for binding? Is that process occurring at the intracellular level?

Allison: No, I think it is just cycling normally.

Hafler: So does capture occur extracellularly and not intracellularly?

Allison: The current working model is that CTLA-4 gets recycled and accumulates in these endosomal particles. It is fired when the TCR is engaged and encounters B7 then.

Hafler: Is B7–1 internalized?

Allison: I don't know.

Healy: Given that *vav* plays a role in capping the TCR (Holsinger et al 1998), have you looked at the effects CTLA-4 may have on capping?

Allison: No, but we need to.

Stockinger: Coming back to the problems in the CTLA-4 knockout mice, did you check how polyclonal the response is? Perhaps they induce responses to bona fide antigens and what's missing is the death by neglect at the end of an immune response because you have unlimited amounts of cytokines around.

Allison: We can't rule it out, but we don't see any perturbation if we stain with a panel of Vβ-specific monoclonal antibodies, so at that level at least it seems to be polyclonal.

Waldmann: On the same theme, if you introduce antigen at the beginning, can you see hijacking of the whole response?

Allison: We haven't done that. What we were trying to do is the reverse experiment. We are trying to restrict the antigen available to a single peptide using the mice made by Phillipe Marrack that express only an ovalbumin peptide covalently linked to IA^d.

Abbas: Remember that those mice are dead at about the time you finish genotyping them, so it's rather difficult to do. They only live for about three weeks.

Allison: As early as five days you can begin to see the cells of activated phenotype.

Abbas: But if you genotype them at about two weeks or so after weaning, then you're in trouble.

Wraith: A direct extension of that question: an important thing to consider before you start doing experiments with Pippa Marrack's mice is what happens if you put them into gnotobiotic conditions. The mice might live more than three weeks.

Abbas: It costs $20 000 per mouse strain to do that!

Allen: Have you tried any of the antigen systems to test this idea that you can lower the threshold by blocking CTLA-4?

Allison: We've been trying this with peptides we got from Mike Bevan. We tried to see whether we could change the dose–response curve. We didn't succeed. He had initially defined partial agonists and antagonists on the basis of sensitization to killing. There was about a 2 log difference. When we did the titration for the ability to induce IL-2 production there was a 5 log difference, and we were unable to make that up by CTLA-4 blockade. More recently, we've got a more tightly related set of peptides in the cytochrome system which we are beginning to look at.

Lechler: We've been wondering for a while what the function of B7 expression on T cells is. In humans you can construct an argument that T cells can present antigen because they also express class II MHC, but in a mouse $CD4^+$ T cell it's difficult to implicate any cognate significance in B7 expression. One does wonder whether CTLA-4/B7 T cell:T cell contact in the context of the cluster around an APC could have a role. The problem with that hypothesis is then you've got B7/CTLA-4 interacting away from the TCR 'cap', because it would be a T:T contact.

Allison: B7 can be relevant. In our first paper in the *Journal of Experimental Medicine* (Krummel & Allison 1995) we showed the levels of B7-2 on T cells would be sufficient to co-stimulate T cells were it not for the attenuation by CTLA-4. With regard to the second part, I don't think that CTLA-4 binding to B7 is going to have any signalling consequences away from the TCR/CD28 plaque.

One could argue that if it was there it would just sop up B7 or something, but what we are beginning to think we know about signalling is that you have to really co-localize the relevant molecules. With the beads, for example, everything has to be on the same bead.

Lenardo: If your model of a strictly quantitative balance between CD28 and CTLA-4 is correct, you might anticipate in the heterozygote that you would get some effect. Do you see that?

Allison: We haven't looked closely enough yet.

Hafler: We have looked at B7-1 and B7-2 to co-stimulation on B cells and we find that in human T cell systems they cannot provide a co-stimulatory signal. It turns out that B7-2 on human activated T cells is hypoglysylated and does not bind CD28 (Höllsberg et al 1997).

Lechler: B7-1 can: we've got pretty good data that T cells can co-stimulate in a primary response.

References

Höllsberg P, Scholz C, Andersen et al 1997 Expression of a hypoglycosylated form of CD86 (B7–2) on human T cells with altered binding properties to CD28 and CTLA-4. J Immunol 15:4799–4805

Holsinger LJ, Swat W, Davidson L, Healy JI, Alt FW, Crabtree GR 1998 Actin cap formation by *vav* in lymphocyte signal transduction, submitted

Krummel MF, Allison JP 1995 CD28 and CTLA-4 have opposing effects on the response of T cells to stimulation. J Exp Med 182:459–465

Antigen-specific CD4$^+$ T cells that survive after the induction of peripheral tolerance possess an intrinsic lymphokine production defect

Kathryn A. Pape, Alex Khoruts, Elizabeth Ingulli, Anna Mondino, Rebecca Merica and Marc K. Jenkins

Department of Microbiology, Center for Immunology, University of Minnesota Medical School, Box 334 FUMC, 420 Delaware Street SE, Minneapolis, MN 55455, USA

Abstract. Injection of soluble foreign antigen without an adjuvant induces a state of antigen-specific immunological unresponsiveness. We investigated the cellular mechanisms that underlie this form of peripheral tolerance by physically tracking a small population of ovalbumin (OVA) peptide/I-A^d-specific, CD4$^+$ T cell receptor (TCR) transgenic T cells following adoptive transfer into normal recipients. Injection of OVA peptide in the absence of adjuvant caused the antigen-specific T cells to proliferate for a brief period after which most of the T cells disappeared. The remaining OVA-specific T cells had converted to a memory phenotype but were poorly responsive *in vivo* as evidenced by a failure to accumulate in the draining lymph nodes following immunization with OVA peptide in adjuvant. These surviving T cells possessed a long-lasting, but reversible, defect in IL-2 and TNF-α production and *in vivo* proliferation, but did not gain the capacity to produce Th2-type cytokines or suppress the clonal expansion of T cells specific for another antigen. Therefore, some antigen-specific T cells survive this peripheral tolerance protocol but are functionally unresponsive due to an intrinsic activation defect.

1998 Immunological tolerance. Wiley, Chichester (Novartis Foundation Symposium 215) p 103–119

Several mechanisms are used by the immune system to prevent self-reactive T cells from responding to self proteins and causing autoimmunity (reviewed by Kruisbeek & Amsen 1996). Developing thymocytes that avidly recognize peptide–major histocompatibility complex (MHC) complexes derived from proteins expressed by antigen-presenting cells (APCs) in the thymus are physically deleted. This process alone, however, cannot account for T cell tolerance that exists toward antigens that are not presented in the thymus, for example, proteins that are expressed exclusively in the parenchymal tissues and proteins that are only expressed at later developmental stages. Tolerance to these types of antigen is particularly problematic because peripheral T cells that patrol extrathymic tissues are designed to respond in a positive fashion unlike immature T cells that are programmed to die when stimulated. Therefore the challenge of the immune system is to retain the capacity to activate peripheral T cells that are specific for microbial antigens, but silence the functions of peripheral T cells that are specific for extrathymic self antigens, without being able to delete these T cells in the thymus.

Clues as to how this could occur can be found in experiments on the immunogenicity of foreign antigens. It has long been known that the T cell immunogenicity of soluble foreign proteins is dependent on the form of antigen administration (reviewed by Janeway 1992). Antigen aggregation and adjuvants promote immunogenicity whereas deaggregation and administration without adjuvant do not. In fact, antigen injection alone often induces unresponsiveness to subsequent challenge with the immunogenic form of the antigen. Because factors that promote immunogenicity (adjuvants, antigen aggregation) do not affect the 'foreign-ness' of the antigen, it is likely that they act by producing the appropriate environment for productive activation of the relevant peptide–MHC-specific T cells. It has been proposed that adjuvants create this environment by inducing the expression of molecules on APCs that deliver essential co-stimulatory signals to T cells that are required for proliferation and differentiation (Janeway 1992). Without these co-stimulatory signals, activated T cells die or become functionally unresponsive (reviewed by Mueller & Jenkins 1995).

Although there is agreement that antigen administration in the absence of adjuvant can induce T cell unresponsiveness, the nature of the cellular mechanism(s) involved is controversial. Injection of peptide antigen without adjuvant into naïve mice causes many of the specific $CD4^+$ T cells to proliferate transiently and then die via apoptosis (Critchfield et al 1994, Liblau et al 1996). However, in most cases not all of the specific T cells are lost and the survivors appear to be functionally unresponsive (at the level of proliferation and IL-2 production) to *in vitro* challenge with the antigen (Kearney et al 1994, Liblau et al 1996). These results suggest that peripheral deletion and functional inactivation contribute to the state of unresponsiveness induced by injection of soluble antigen. However, other studies indicate that the surviving T cells are not globally unresponsive, and in fact produce high levels of Th2 lymphokines, for example IL-4 (Burstein & Abbas 1993). Proponents of this 'immune deviation' theory argue that injection of antigen without adjuvants does not really induce tolerance, just a different type of immune response that does not rely on T cell proliferation or IL-2 production (Rocken & Shevach 1996). Furthermore, it has been proposed that the 'deviated' T cells produce cytokines such as TGF-β that suppress the activation of other T cells in a bystander fashion (Chen et al 1994).

Development of the adoptive transfer model system

We reasoned that a method that allowed physical detection of an antigen-specific $CD4^+$ T cell population would be required to assess the involvement of the various mechanisms mentioned above in the unresponsive state induced by injection of soluble antigen. This is difficult to do for technical reasons because the frequency of T cells specific for any given antigenic peptide is exceedingly low; at most 1/1000 T cells in immunized individuals. We attempted to get around this clonal infrequency problem by using TCR transgenic mice in which the majority of T cells are specific for a single peptide–MHC complex. Initial attempts were made to induce peripheral tolerance in the DO11.10 transgenic mice that contain in their germline DNA, rearranged TCR α and β genes that encode a TCR specific for chicken ovalbumin 323–339 (hereafter referred to as OVA peptide) bound to I-A^d class II MHC molecules (Murphy et al 1990). The transgenic TCR can be detected with the KJ1-26 monoclonal antibody (mAb) that binds only to this particular TCR heterodimer (Haskins et al 1983). Because the transgenes inhibit rearrangement of endogenous TCR genes and because the transgenic TCR is class II MHC-restricted, a large fraction of the T cells in these mice are positively selected in the thymus to become $CD4^+$, KJ1-26^+ peripheral T cells.

Unfortunately, the TCR transgenic mice responded differently to peptide injection than normal mice. It was difficult to demonstrate any improvement or decrement in the vigorous *in vitro* proliferation of DO11.10 T cells in response to OVA following priming of the transgenic mice with OVA plus adjuvant or tolerization with OVA alone (Kearney et al 1994). These results suggested that the TCR transgenic mice do not mount normal immune responses to antigenic challenge *in vivo*, presumably because of the artificially high frequency of antigen-specific cells present. Forster et al (1995) and Lanoue et al (1997) later reported the similar finding that peripheral tolerance mechanisms could be overwhelmed in TCR transgenic mice that expressed a very high number of self-reactive T cells. However the high frequency of antigen-specific T cells is not an absolute barrier to the study of peripheral tolerance, because some investigators have been able to induce T cell unresponsiveness with soluble antigen injections in certain TCR transgenic mouse lines (Liblau et 1996, Falb et al 1996).

To get around the difficulties encountered in the intact TCR transgenic mice, an adoptive transfer system was established where the TCR transgenic T cells were not the dominant T cell population, but could still be physically detected using the KJ1-26 mAb that binds exclusively to DO11.10 T cells (Kearney et al 1994). After transfer into unirradiated BALB/c mice, a small population of phenotypically naïve (as defined by high expression of CD45RB; Kearney et al 1995) $CD4^+$, KJ1-26^+ cells (routinely 0.2–0.6% of lymph node cells) was detected in the recipient's lymph nodes. The naïve DO11.10 T cells persisted in unimmunized recipients with a half-life of about two weeks. The cells did not divide or produce lymphokines *in vivo*, and retained a naïve surface phenotype at all times after transfer as long as OVA was not present.

Early events associated with the induction of peripheral tolerance

The adoptive transfer system was used to monitor the behaviour of the transferred DO11.10 T cells following injection of OVA peptide without adjuvant, a protocol known to induce T cell unresponsiveness (Janeway 1992). Because adjuvants antagonize tolerance induction and promote immunity (Janeway 1992), the behaviour of the DO11.10 cells following injection of OVA with the adjuvants lipopolysaccharide (LPS) or complete Freund's adjuvant (CFA) was also studied. As early as 4 h after injection of intact OVA or OVA peptide alone, IL-2-producing DO11.10 T cells were detected in the lymph nodes as assessed by *ex vivo* intracellular IL-2 staining (A. Khoruts, A. Mondino & M. Jenkins, unpublished work 1997). *In vivo* IL-2 production by the DO11.10 T cells peaked at 10 h and faded rapidly thereafter. Injection of OVA peptide with adjuvant increased by three- to fourfold the number of IL-2-producing DO11.10 T cells present at 10 h when compared with injection of OVA peptide alone. Staining of lymph node sections with anti-CD11c mAb revealed that many of the DO11.10 T cells from mice injected with OVA were found in physical proximity to dendritic cells within the paracortex (T cell-rich region) at the same time that IL-2-producing DO11.10 T cells were identified (Ingulli et al 1997). The simplest explanation for this result is that dendritic cells are the first APCs to present peptides from soluble antigens to specific T cells.

In vitro experiments indicate that co-stimulatory signals through CD28 play an important role in IL-2 production by activated T cells (Lenschow et al 1996). To test this in our system, we transferred CD28-deficient DO11.10 T cells into normal recipients, which were then injected with OVA with or without adjuvant. The adoptively-transferred CD28-deficient DO11.10 T cells made very little IL-2 *in vivo* at any time following intravenous injection of OVA alone and the presence of adjuvant did not enhance production as it did for normal DO11.10 T cells (A. Khoruts, A. Mondino, S. Reiner & M. Jenkins, unpublished work 1997). Therefore, CD28-mediated co-stimulation was important for achieving maximal IL-2 production *in vivo*, and for the ability of adjuvants to enhance IL-2 production. As discussed below, these results are consistent with the possibility that adjuvants increase the expression of the CD28 ligands B7-1 and B7-2 on APCs.

By 48–72 h, the number of DO11.10 T cells present in the lymph nodes increased about 10-fold over the starting level following injection of OVA alone and about 40-fold following injection of OVA plus adjuvant (Kearney et al 1994). The adjuvant-dependent increase in clonal expansion is likely to be related to the aforementioned increase in the production of IL-2 or some other T cell growth factor. Bromodeoxyuridine labelling experiments showed that most of the DO11.10 T cells proliferated during this period whether adjuvant was present or not (Pape et al 1997a). Therefore, the increased number of DO11.10 T cells that accumulated in the lymph nodes when adjuvant was present is probably explained by a quantitative, not a qualitative difference in the number of times the cells divided. The DO11.10 T cells also converted to an activated surface phenotype (CD45RB^{low}, CD62L^{low}, LFA-1^{high}) during this period and again this change occurred in response to OVA injection with or without adjuvant. Interestingly, the capacity of LPS and CFA to enhance clonal expansion and follicular entry of antigen-stimulated T cells was mimicked by injection of recombinant IL-1α, suggesting that adjuvanticity is related in some way to inflammation (Pape et al 1997b).

Following the peak of accumulation at 48–72 h, many of the DO11.10 T cells disappeared from the lymphoid tissues (Kearney et al 1994) as described in several other systems (Webb et al 1990, Rocha & von Boehmer 1991). This loss of cells was particularly dramatic following injection of OVA alone where the number of DO11.10 T cells remaining in the lymphoid tissues fell to the input level by day 7. In contrast, DO11.10 T cells were present in the lymphoid tissues at a level 10-fold higher than the starting level 7 d after injection of OVA plus adjuvant. The loss of antigen-specific T cells could be related to cell death. Although *ex vivo* flow cytometric analyses of lymph node cells failed to identify an increase in apoptotic DO11.10 T cells during the decline phase of the response, DO11.10 T cells removed from mice 72 h after injection of OVA alone died rapidly *in vitro* (A. Mondino, unpublished work 1997). Therefore, apoptosis may be occurring during the decline phase of the response, but apoptotic cells may be rapidly disposed of *in vivo* via a pathway (macrophage phagocytosis) that does not operate *in vitro*.

The DO11.10 T cells that survived 1–2 weeks after OVA injection still expressed a memory surface phenotype suggesting that these cells had been stimulated by antigen in the past. These DO11.10 T cells produced very little IL-2 or TNF-α *in vivo* 6–10 h after rechallenge with OVA (K. Pape, A. Khoruts, A. Mondino & M. Jenkins, unpublished work 1997), and proliferated poorly in the lymphoid tissues at later times (Kearney et al 1994). If mice were rechallenged with OVA seven weeks after tolerization with OVA alone, then the DO11.10 T cells proliferated vigorously *in vivo*, indicating that the functional unresponsiveness was reversible (K. Pape & M. Jenkins, unpublished work 1997). DO11.10 T cells that remained several weeks after injection of OVA plus adjuvant produced large amounts of IL-2 and TNF-α *in vivo*, although these primed cells produced the lymphokines more rapidly than naïve cells (K. Pape, A. Khoruts, A. Mondino, & M. Jenkins, unpublished work 1997). Therefore, although the DO11.10 T cells that survived the injection of OVA alone or OVA plus adjuvant both possessed the memory cell surface phenotype, the latter population was defective with respect to IL-2 and TNF-α production *in vivo*, whereas the former population produced these cytokines with the enhanced kinetics of memory cells. DO11.10 T cells from naïve, OVA-, or OVA plus adjuvant-injected mice did not produce detectable levels of the Th2 lymphokines, IL-4 or IL-5, in response *in vitro* rechallenge with OVA (K. Pape & M. Jenkins, unpublished work 1997). Thus, no evidence of immune deviation was obtained.

The impaired capacity of DO 11.10 T cells in tolerized mice to undergo clonal expansion in response to *in vivo* rechallenge provided us with an opportunity to test for bystander suppression. Normal mice received DO11.10 T cells and either no injection or an intravenous injection of OVA. Two weeks earlier, these mice were injected with a second population of TCR transgenic T cells specific for a haemagglutinin peptide–I-E^d complex (Kirberg et al 1994). This second population of T cells was tracked with another anti-clonotypic mAb called 6.5. Shortly after transfer of the haemagglutinin peptide-specific T cells, the mice were immunized with a mixture of the OVA and haemagglutinin peptides in CFA. The clonal expansion of each T cell population was quantified by flow cytometric analyses of draining lymph node cells stained with the anti-clonotypic mAbs. The haemagglutinin peptide-specific T cells accumulated to the same extent in lymph nodes that contained no DO11.10 T cells, or naïve or tolerized DO11.10 T cells. These results suggest that tolerized DO11.10 T cells do not suppress the activation of bystander T cells within the same lymphoid tissue.

Model

The models shown in Figs 1 and 2 are based on our comparison of the behaviour of adoptively-transferred antigen-specific CD4$^+$ T cells under conditions where injection of antigen alone results in peripheral tolerance (Fig. 1) or where adjuvants convert the response to immunity (Fig. 2). A key tenet of this model is that APC activation determines whether antigen presentation to specific T cells results in immunity or tolerance.

On the basis of the co-localization of DO11.10 T cells and paracortical dendritic cells shortly after OVA injection, we postulate that the dendritic cells that reside in the T cell areas of lymphoid tissues take up, process and present OVA peptide–I-A^d complexes soon after systemic injection of OVA. These dendritic cells have been shown by Inaba et al (1994) to be in a non-activated state as evidenced by a lack of CD40 and B7-1 expression and low level expression of B7-2. We propose that naïve DO11.10 T cells then interact with these resting dendritic cells, which present OVA peptide–I-A^d complexes and some B7-2 molecules. This stimulates the T cells to produce a small amount of T cell growth factor, perhaps IL-2, and

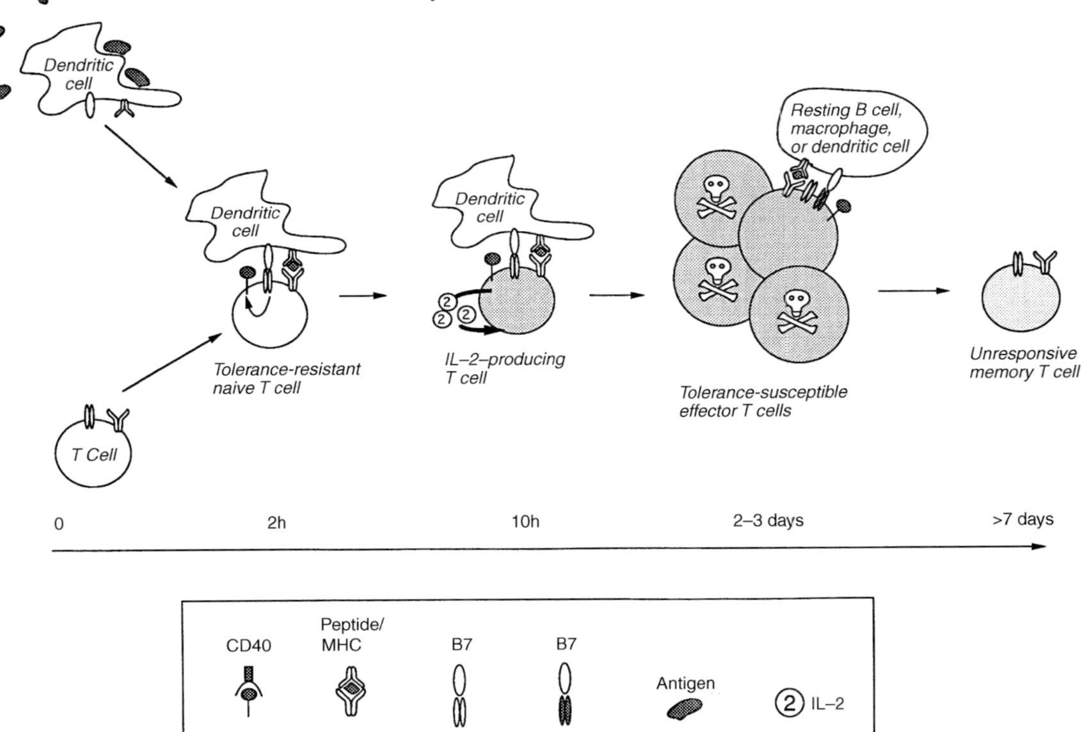

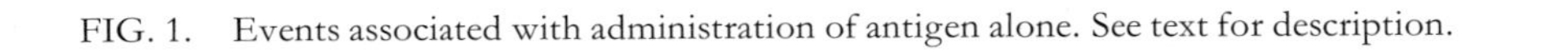

FIG. 1. Events associated with administration of antigen alone. See text for description.

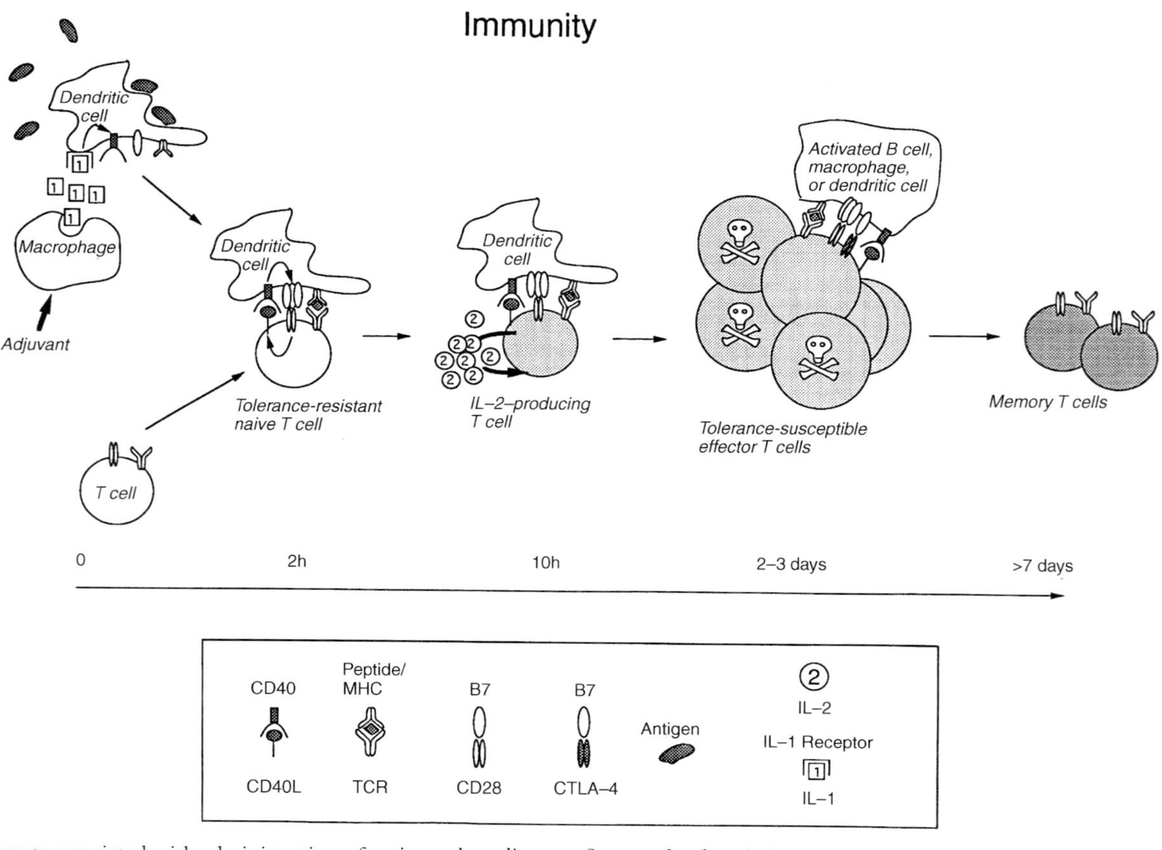

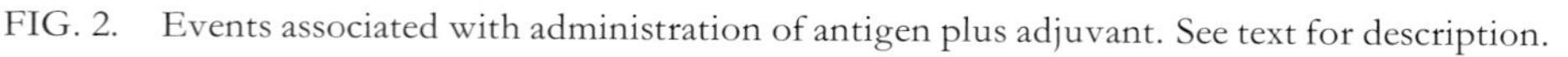

FIG. 2. Events associated with administration of antigen plus adjuvant. See text for description.

proliferate briefly. The finding of Perez et al (1997) that blockage of B7 molecules at the time of soluble antigen injection prevents tolerance induction indicates that CD28-mediated activation signals are required for naïve T cells to enter a 'tolerance-susceptible' state. Because adjuvant effects are absent when antigen alone is injected, B7 molecule expression on dendritic cells, macrophages and B cells is not up-regulated, and continued presentation of OVA peptide–I-A^d complexes by these APCs to the 'tolerance-susceptible' T cells causes most of these cells to die and the survivors to become functionally unresponsive. These outcomes could be explained by gain of the capacity to produce 'anergy proteins' (Jenkins 1992) or by induction of inhibitory signalling molecules such as CTLA-4 (Walunas et al 1994). The report of Perez et al (1997) that blockage of CTLA-4 at the time of injection of soluble OVA prevents the induction of functional unresponsiveness is consistent with the latter possibility. CTLA-4, which is rapidly induced on activated T cells, binds to B7 molecules with a much higher affinity than CD28 (Linsley et al 1991). Therefore it is conceivable that under conditions of low-level B7-2 expression by resting dendritic cells, CTLA-4 would be preferentially bound on interacting OVA peptide-specific T cells, resulting in the functional inactivation of the T cells. An interesting implication of this model first proposed by Finkelman et al (1996) is that dendritic cells, thought to be the champion APC for activating naïve T cells, may be tolerizing APCs unless they are activated.

If an adjuvant is present, then we postulate that B7 molecules will be induced above the basal level. LPS has been shown to directly up-regulate B7 expression on B cells and macrophages *in vitro* and on resting dendritic cells *in vivo*. Another adjuvant-dependent mechanism of B7 up-regulation may operate via CD40 signalling in APCs. Adjuvants would be expected to stimulate macrophages to produce IL-1, which has been shown to induce CD40 molecules on dendritic cells *in vitro* (McLellan et al 1996). The CD40 molecules could then be bound by CD40 ligand expressed on activated T cells to which the dendritic cell is presenting antigen. CD40 signalling has been shown to up-regulate B7 expression on APCs (McLellan et al 1996). B7 up-regulation on the dendritic cells and other APCs would result in increased CD28 signalling in antigen-specific T cells, which would cause increased production of T cell growth factors and survival proteins. Adjuvant-dependent induction of high levels of B7 expression on APCs may allow CD28 to out-compete less abundant CTLA-4 molecules for B7 binding, thus ensuring that the positive signals mediated by CD28 outweigh the negative signals transduced by CTLA-4. The net effect would be increased proliferation and survival of antigen-stimulated T cells. It may be that the increased amount of proliferation caused by adjuvants prevents antigen-stimulated T cells from becoming unresponsive, as suggested by *in vitro* experiments with T cell clones (Jenkins 1992).

The molecular basis for the IL-2 and TNF-α production defects exhibited by the DO11.10 T cells that survive after injection of soluble OVA peptide is unknown. These T cells have normal numbers of surface CD4 molecules but only about one-half the number of surface TCR molecules expressed by naïve T cells (K. Pape & M. Jenkins, unpublished work 1997). However, it is difficult to believe that this reduction is functionally significant given the low number of surface TCRs that need to be engaged for signalling to occur. We think it is more probable that the lymphokine production defects are secondary to proximal TCR signalling defects, that may be similar to those observed in anergic Th1 clones. These cells fail to produce IL-2 because of a defect in production of the AP-1 transcription factor that is critical for TCR-dependent IL-2 mRNA expression (Mondino et al 1996). Future experiments will be required to determine if DO11.10 T cells that become unresponsive *in vivo* have a similar defect.

Acknowledgements

This work was supported by NIH grants AI27998, AI35296 and AI39614. A. Khoruts is a Howard Hughes Medical Institute Physician Postdoctoral Fellow and a recipient of the GIDH Basic Research Award. K. Pape and R. Merica were supported by NIH training grants AI07313 and AI07421, respectively.

References

Burstein HJ, Abbas AK 1993 *In vivo* role of IL-4 in T cell tolerance induced by aqueous protein antigen. J Exp Med 177:457–463

Chen Y, Kuchroo VK, Inobe J, Hafler DA, Weiner HL 1994 Regulatory T cell clones induced by oral tolerance: suppression of autoimmune encephalomyelitis. Science 265:1237–1240

Critchfield JM, Racke MK, Zuniga-Pflucker JC et al 1994 T cell deletion in high antigen dose therapy of autoimmune encephalomyelitis. Science 263:1139–1143

Falb D, Briner TJ, Sunshine GH et al 1996 Peripheral tolerance in T cell receptor-transgenic mice: evidence for T cell anergy. Eur J Immunol 26:130–135

Finkelman FD, Lees A, Birnbaum R, Gause WC, Morris SC 1996 Dendritic cells can present antigen *in vivo* in a tolerogenic or immunogenic fashion. J Immunol 157:1406–1414

Forster I, Hirose R, Arbeit JM, Clausen BE, Hanahan D 1995 Limited capacity for tolerization of CD4+ T cells specific for a pancreatic beta cell neo-antigen. Immunity 2:573–585

Haskins K, Kubo R, White J, Pigeon M, Kappler J, Marrack P 1983 The MHC-restricted antigen receptor on T cells. I. Isolation of a monoclonal antibody. J Exp Med 157:1149–1169

Inaba K, Witmer-Pack M, Inaba M et al 1994 The tissue distribution of the B7-2 costimulator in mice: abundant expression on dendritic cells *in situ* and during maturation *in vitro*. J Exp Med 180:1849–1860

Ingulli E, Mondino A, Khoruts A, Jenkins MK 1997 *In vivo* detection of dendritic cell antigen-presentation to CD4+ T cells. J Exp Med 185:2133–2141

Janeway CA 1992 The immune system evolved to discriminate infectious nonself from noninfectious self. Immunol Today 13:11–16

Jenkins MK 1992 The role of cell division in the induction of clonal anergy. Immunol Today 13:69–73

Kearney ER, Pape KA, Loh DY, Jenkins MK 1994 Visualization of peptide-specific T cell immunity and peripheral tolerance induction *in vivo*. Immunity 1:327–339

Kearney ER, Walunas TL, Karr RW et al 1995 Antigen-dependent clonal expansion of a trace population of antigen-specific CD4+ T cells *in vivo* is dependent on CD28 costimulation and inhibited by CTLA-4. J Immunol 155:1032–1036

Kirberg J, Baron A, Jakob S, Rolink A, Karjalainen K, von Boehmer H 1994 Thymic selection of CD8+ single positive cells with a class II major histocompatibility complex-restricted receptor. J Exp Med 180:25–34

Kruisbeek AM, Amsen D 1996 Mechanisms underlying T-cell tolerance. Curr Opin Immunol 8:233–244

Lanoue A, Bona C, von Boehmer H, Sarukhan A 1997 Conditions that induce tolerance in mature CD4+ T cells. J Exp Med 185:405–414

Lenschow DJ, Walunas TL, Bluestone JA 1996 CD28/B7 system of T cell costimulation. Annu Rev Immunol 14:233–258

Liblau RS, Tisch R, Shokat K et al 1996 Intravenous injection of soluble antigen induces thymic and peripheral T-cells apoptosis. Proc Natl Acad Sci USA 93:3031–3036

Linsley PS, Brady W, Urnes M, Grosmaire LS, Damle NK, Ledbetter JA 1991 CTLA-4 is a second receptor for the B cell activation antigen B7. J Exp Med 174:561–569

McLellan AD, Sorg RV, Williams LA, Hart DN 1996 Human dendritic cells activate T lymphocytes via a CD40:CD40 ligand-dependent pathway. Eur J Immunol 26:1204–1210

Mondino A, Whaley CD, DeSilva DR, Li W, Jenkins MK, Mueller DL 1996 Defective transcription of the IL-2 gene is associated with impaired expression of c-Fos, FosB and JunB in anergic Th1 cells. J Immunol 157:2048–2057

Mueller DL, Jenkins MK 1995 Molecular mechanisms underlying functional T-cell unresponsiveness. Curr Opin Immunol 7:375–381

Murphy KM, Heimberger AB, Loh DY 1990 Induction by antigen of intrathymic apoptosis of $CD4^{+}CD8^{+}TCR^{lo}$ thymocytes *in vivo*. Science 250:1720–1723

Pape KA, Kearney ER, Khoruts A et al 1997a Use of adoptive transfer of T cell antigen receptor transgenic T cells for the study of T cell activation *in vivo*. Immunol Rev 156:67–78

Pape KA, Kearney ER, Mondino A et al 1997b Inflammatory cytokines enhance the *in vivo* clonal expansion and differentiation of antigen-activated $CD4^{+}$ T cells. J Immunol 159:591–598

Perez V, van Parijs L, Biuckians A, Zheng XX, Strom TB, Abbas AK 1997 Induction of peripheral tolerance *in vivo* requires CTLA-4 engagement. Immunity 6:411–418

Rocha, B, von Boehmer H 1991 Peripheral selection of the T cell repertoire. Science 251:1225–1228

Rocken M, Shevach EM 1996 Immune deviation — the third dimension of nondeletional T cell tolerance. Immunol Rev 49:175–194

Walunas TL, Lenschow DJ, Baker CY et al 1994 CTLA-4 can function as a negative regulator of T cell activation. Immunity 1:405–413

Webb S, Morris C, Sprent J 1990 Extrathymic tolerance of mature T cells: clonal elimination as a consequence of immunity. Cell 63:1249–1256

DISCUSSION

Waldmann: I'm aware that this might be impossible, but have you done any sort of experiment where you've linked the two peptides together in some form? In other words, have you tried to look for linked suppression?

Jenkins: In the experiment I presented the epitopes were not linked; they were just mixed in the adjuvant and delivered together. We have not done the experiment of making the linked peptide. We went back and looked at some of the previous experiments of Weiner et al on bystander suppression, and found that linkage would not be expected to be necessary if the peptides are delivered to the same anatomical location.

Waldmann: Although Weiner's interpretation would be the effects of suppression are operating in the periphery, presumably in the tissue.

Jenkins: But those effects could be read out in the draining lymph nodes in his experiments.

Hafler: The key to those experiments is that you have one population that secretes either IL-4 or TGF-β.

Jenkins: Right, and we have not been able to show that the DO11.10 T cells that survive after i.v. injection of antigen in our system make either of those.

Hafler: Exactly. So you wouldn't predict so-called 'bystander suppression'.

Mitchison: That's an ingenious experiment, but I don't quite see why you need to do it. Why can't you just do it entirely with OVA-specific cells?

Jenkins: We could, but we would need a way to track the two different populations simultaneously. There are ways to do this, but we haven't tried them yet.

Shevach: Let me tell you how to do the experiment, which is a positive control for bystander immune deviation. You could stimulate one of the populations *in vitro* under Th2 conditions, transfer the cells and then doubly prime the recipient.

Abbas: That is a difficult experiment to do because they no longer go to the lymph nodes, and if you recover T cells from lymph nodes you get Th1s coming back.

Jenkins: We've done that experiment and that's exactly what happens. When we activate them *in vitro* and do the transfer we can't find them in the lymphoid tissue, which we think is because they lose expression of CD62L (L-selectin).

Abbas: Two related questions. If you give i.v. peptide and LPS do you prevent tolerance induction? The second is the flipside of that: if you prime your CD28 knockout T cells with peptide in adjuvant subcutaneously and then come back later, do they get tolerized?

Jenkins: We've addressed the first question but not the second. If you give OVA with LPS you will get what is essentially a systemic adjuvant effect. In every lymph node in the animal's body the clonal expansion is enhanced, the T cells go into the follicles and at least on the basis of *in vitro* cytokine production they are now primed cells, so they make interferon (IFN)-γ whereas the naïve cells don't make IFN-γ.

Abbas: The adoptive transfer of the CD28 knockout, which we are also in the processes of doing, is obviously a nice way of asking about CTLA-4 effects in the absence of CD28 *in vivo*. This is probably the best way to do that experiment.

Jenkins: When the CD28-deficient DO11.10 T cells are transferred into a normal mouse they have to compete with T cells that have CD28. In this situation, their clonal expansion in response to antigen is so low that it is difficult to imagine doing a tolerance experiment, where an even lower response would have to be detected.

Kioussis: One thing that is of note in the peptide-injection experiment, and also in experiments with superantigen, is the disappearance of reactive cells 24 h after administration of the antigen. Do the numbers go down in your system as well, and where do the cells go?

Jenkins: Particularly 24 h after injection, the number of cells goes down in the lymphoid tissues. However, if you monitor with immunohistology rather than flow cytometry, this decrease is not detected. Our interpretation of this is that the cells are highly activated and are in fact sticking to the stromal elements or to dendritic cells that are sticking to stromal elements. We think the activated T cells remain attached to the lymph node debris that most of us normally filter out of single cell suspensions.

Kioussis: We see the same thing. When we look with immunohistology the cells are there. However, we cannot see them in fluorescence-activated cell sorting (FACS) analysis.

Jenkins: We're trying to figure out a way to release the activated T cells from the stroma, because 24 h is an interesting time point for analysis of early activation events.

Stockinger: I am not convinced that you can call these cells 'tolerant' on the basis of this difference after re-stimulation, which is not much more than a blip in the other population. Would it be possible to assume that administration of peptide without adjuvants gives an overriding CTLA-4-driven response?

Jenkins: I think that the CTLA-4 finding is a real breakthrough in the area. If you look at the dendritic cells that are in the lymphoid tissue at the time antigen is given without adjuvant, you find that they have a little bit of B7-2. This is exactly the situation where CTLA-4 could out-compete CD28 for B7 because of its high affinity and cause suppression of T cell function. But when the adjuvant is there, B7 is induced and now the more abundant CD28 wins out, allowing enhancement of T cell function.

Stockinger: Following on from that, perhaps they are not tolerant, but rather they are memory cells. You give them very little time: you re-challenge after 10 days. What would happen if you re-challenged after four weeks?

Jenkins: If we re-challenge four weeks after injection of soluble antigen, then the transferred DO11.10 T cells are still unresponsive *in vivo*. However, if the interval is extended to seven weeks, the T cells respond to *in vivo* antigenic challenge as well as naïve T cells. We've done an assay where we give the antigen i.v., transfer in the DO11.10 T cells and then ask whether the transferred cells become blasts as would be expected if the peptide–MHC complex is still in the recipient's body. On the basis of this assay, it appears that the OVA peptide–MHC complex disappears about a week after injection. Work by others suggests that persistent TCR stimulation is required to maintain the unresponsive state and thus the cells probably recover because the peptide–MHC complex disappears.

Mitchison: Surely the brief duration that you observe argues against peripheral tolerance of the Dresser–Weigle type that was referred to in your talk. It has long been thought that their type of tolerance is of long duration in adults, with a timing of recovery after administration of antigen stops that is thought to reflect the kinetics of new T cells emerging from the thymus rather than the loss of stored antigen (Mitchison 1962, Taylor 1968). That conclusion is supported by the observations that the rate of recovery varies sharply with age (in the chicken), and that thymectomy post-tolerization prevents recovery.

Jenkins: The seven week period required for the DO11.10 T cells to recover their responsiveness in our experiments fits with the time required for new T cells to be produced in the thymus and exported to the periphery. However, although we have not done the thymectomy experiment, the fact that the DO11.10 T cells cannot be produced by the recipient's thymus at a detectable frequency, argues that repopulation of the periphery with functional new thymic immigrants cannot explain the recovery we observed in our experiments.

Arnold: Are the differences in the follicular migration due to differential expression of adhesion molecules? In other words, do the mycobacteria which you inject with the adjuvant increase certain adhesion molecules and therefore you see different migration?

Jenkins: That is a great question. When the cells get activated, whether the adjuvant is there or not, they convert to the classic memory phenotype: they lose CD45RB and CD62L, and LFA-1 goes up. We haven't found a marker that differentiates the two situations.

Arnold: I'm asking about the molecules on the tissues, not on the T cells.

Jenkins: I think Blr-1 is a great candidate there. Blr-1 is an orphan chemokine receptor that, when knocked out, causes B cells to lose their way to the follicles. A simple model would be that some cell in the follicle such as a follicular dendritic cell secretes the Blr-1 ligand, and then any cell that has that receptor will go in that direction. It has been shown that resting T cells lack Blr-1 (Förster et al 1996). If immunization with antigen and adjuvant causes T cells to induce Blr-1, then they might do what B cells do — go into follicles. We're having a hard time getting hold of the reagent required to test this, but we would love to do that experiment.

Kurts: Your experiments clearly suggest that something happens to the APC when you inject the adjuvant: it gives a different signal to the responding T cell. You suggested that interleukin production is one of the mediators. What about up-regulation of surface molecules such as CD40? Have you any information that these are differentially expressed?

What happens to the APCs after the initial cluster formation? They seemed to disappear. Are they killed?

Jenkins: Yes, CD40 goes up when you give LPS. The resting dendritic cells of secondary lymphoid tissues don't have any CD40. Clearly CD40 signalling in dendritic cells causes them to make IL-12, so I think that is going to be very important. Interestingly, in a culture system, Hart and co-workers have shown that addition of IL-1 and TNF-α will induce CD40 on cultured dendritic cells (McLellan et al 1996). A nice loop there might be that adjuvant stimulates IL-1, IL-1 induces CD40, and CD40 is known to be involved in regulation of B7. This chain of events that would give you a really good APC.

With regards to what happens to the APC in a situation where we transfer dye-labelled T cells and dendritic cells that we can track at the same time, early in the response the T cells cluster on the dendritic cell but later the dendritic cell disappears. This either means the T cell kills the dendritic cell, or the dendritic cell leaves the lymph node, or gets activated and metabolizes the dye. It's interesting to think that the T cells could initially benefit from the APC and then kill it.

Kurts: What is the time frame?

Jenkins: 24 h is the peak of cluster formation, and by 48 h the number of dendritic cells is reduced by about 75%.

Healy: How much of your adjuvant effect is accounted for by IL-1? Have you done blocking studies to show that IL-1 really is responsible?

Jenkins: We've failed to block the effect of LPS by injecting anti-TNF-α or anti-IL-1 receptor. Thus, we find ourselves in the situation of not knowing whether we failed to neutralize all of the relevant cytokine or whether there's redundancy in the system.

Mitchison: But is IL-1 active?

Jenkins: Yes. When OVA plus IL-1 is injected, the T cell behaviour is indistinguishable from OVA plus LPS, in terms of the rise and fall in T cell number and the follicular migration. There are some differences in the cytokines that the T cells make: OVA plus IL-1 injection does not result in IFN-γ-producing T cells, whereas injection of OVA plus LPS does. This is probably explained by the fact that LPS, but not IL-1, induces IL-12 production by macrophages.

Mitchison: It's a funny business, because IL-1 has been reported to be the only cytokine which will do that (Finkelman et al 1996).

Jenkins: In our blocking experiments we are asking a lot to be able to sop up all the IL-1 receptor in a mouse stimulated with LPS, so there could be a trivial explanation for our results.

Waldmann: I want to address the dendritic cell side of things. Barbara Fazekas thinks of dendritic cells as being of two types, lymphoid and myeloid. Her idea of the adjuvant effect is that when you have an adjuvant the myeloid cells migrate into the T cell area. But when you get presentation by the lymphoid cells, they only present for tolerance. Is the marker you're using for your dendritic cells a myeloid marker?

Jenkins: N418 marks the majority of both populations as far as I know. We don't know whether the dendritic cells that we found T cells interacting with were resident or migrated from the skin, for instance. In the lymph node there is not an obvious marginal zone equivalent for them to come from the way that there is in the spleen.

Mitchison: Incidentally, Barbara Fazekas says that she can see an effect of priming in the intact anti-OVA TCR transgenic mouse.

Jenkins: I don't want to push that too far. I think our initial attempts in the intact TCR transgenic mice weren't that clever: we didn't thymectomize and we could have done much more sophisticated experiments, such as looking for changes in activation markers.

Hämmerling: Are you suggesting that dendritic cells that have not been stimulated properly by adjuvant will not activate T cells?

Jenkins: Yes. They will stimulate this small abortive T cell response, which is CD28 and B7 dependent. Only when adjuvants increase B7 expression on dendritic cells will they stimulate a productive T cell response.

Hämmerling: All your experiments are supportive of this idea, but is there more direct evidence?

Jenkins: We have no direct evidence. To answer this we would really need a dendritic cell knockout mouse. I think Fred Finkelman has done a nice experiment in this area (Finkelman et al 1996). He looked at the T cell response to peptides from the Fc portions of antibodies. He has found that if he uses a dendritic cell-specific antibody that targets that Fc to dendritic cells *in vivo*, he will actually get tolerance against that epitope. So antigen targeting to dendritic cells gives tolerance. If he gives that antibody plus IL-1 he actually gets priming.

Stockinger: Both in the C5 transgenic system and also in Harald von Boehmer's mice, if you transfer T cells into an intact unmanipulated animal where the antigen is present, the T cells expand massively and are activated. There's no need in that system to trigger the dendritic cells, yet it is dendritic cell presentation (these are Rag mice). To say that dendritic cells will only activate when they get IL-1 is an overinterpretation of your results.

Jenkins: Remember that in the data I showed there is a quantitative difference in the clonal expansion. When adjuvant is not given, clonal expansion still occurs; the adjuvant simply makes it better. I think in Harald's case, the appropriate experiment would be to transfer cells in the presence or absence of adjuvant. He only tested clonal expansion in the absence of adjuvant, so we do not know whether adjuvant would have improved the clonal expression.

Kamradt: With regard to your blip in numbers, do you have any way of differentiating whether the reduction is due to cell death or just migration?

Jenkins: Clearly, cells are leaving the lymph nodes because they enter the efferent lymph. I suspect that cells are also dying. Lenardo and co-workers and McDevitt and co-workers did experiments where they gave a peptide antigen to intact T cell transgenic mice and saw apoptosis of the T cells. We're transferring small numbers of cells and I think that any dying cells are very efficiently disposed of by macrophages. We've had a hard time putting together TUNEL methods with immunohistochemical detection of the transgenic TCR, so we have not been able to show that a particular dying cell is one of the ones we put in.

Mason: Could I answer Herman Waldmann's question about the bystander suppression, at least in another *in vivo* system. We've been using the encephalitogenic peptide of myelin basic protein (MBP), and we can induce non-responsiveness to that and the animals don't get experimental autoimmune encephalomyelitis (EAE). If we then challenge them with whole MBP in Freund's adjuvant, they still don't get disease, but if we take the cells out they won't respond to the peptide but they will respond to the whole protein. So there's no evidence that the response to the whole protein has been modified by the fact that we're induced non-responsiveness to just one peptide.

References

Finkelman FD, Lees A, Birnbaum R, Gause WC, Morris SC 1996 Dendritic cells can present antigen *in vivo* in a tolerogenic or immunogenic fashion. J Immunol 157:1406–1414

Förster R, Mattis AE, Kremmer E, Wolf E, Brem G, Lipp M 1996 A putative chemokine receptor, BLR1, directs B cell migration to defined lymphoid organs and specific anatomic compartments of the spleen. Cell 87:1037–1047

McLellan AD, Sorg RV, Williams LA, Hart DN 1996 Human dendritic cells activate T lymphocytes via a CD40:CD40 ligand-dependent pathway. Eur J Immunol 26:1204–1210

Mitchison NA 1962 Tolerance of erythrocytes in poultry: loss and abolition. Immunology 55:359–369

Taylor RB 1968 Immune paralysis of thymus tissue cells by bovine serum albumin. Nature 220:611

Antigen-specific tolerance induction and the immunotherapy of experimental autoimmune disease

S. M. Anderton, C. Burkhart, G. Y. Liu*, B. Metzler and D. C. Wraith

*Department of Pathology and Microbiology, University of Bristol, School of Medical Sciences, Bristol BS8 1TD and *Department of Pathology, University of Cambridge, Tennis Court Road, Cambridge, CB2 1QP, UK*

Abstract. Antigen-specific tolerance induction is the ultimate goal for specific immunotherapy of autoimmune diseases. Here we will discuss recent experiments designed to induce tolerance following mucosal administration of antigens in a mouse model of experimental autoimmune encephalomyelitis (EAE). We were unable to induce oral tolerance either with whole myelin, myelin basic protein (MBP) or the immunodominant peptide antigen. Oral tolerance was possible, however, with an analogue of the immunodominant peptide modified to increase its affinity for the restricting major histocompatibility complex (MHC) antigen. By contrast, intranasal deposition of peptide antigen proved highly effective for both prevention and treatment of EAE. Prevention of disease was directly related to the antigenic property of the peptide which, in itself, was related to affinity for MHC. Notably, administration of a single peptide was shown to inhibit disease involving multiple epitopes. We investigated the resulting bystander regulation by studying the cellular basis of peripheral tolerance in a transgenic model. These studies indicate that bystander regulation may be the consequence of selective cytokine secretion.

1998 Immunological tolerance. Wiley, Chichester (Novartis Foundation Symposium 215) p 120–136

Current therapeutic strategies for treatment of autoimmune diseases such as multiple sclerosis (MS) are far from satisfactory. They include drugs that non-specifically target proliferating cells but are associated with unacceptable levels of toxicity and drugs targeted to inflammatory processes such as corticosteroids which, while relatively effective, are also associated with undesirable side-effects. New drugs in clinical trial include antibodies directed to cell surface markers on T lymphocytes (CD4, CDW52) and cytokines, such as interferons (IFNs) α and β, that are known to regulate inflammatory responses (Wraith 1998). Sadly, none of these therapeutic approaches specifically targets the cellular mechanisms involved in autoimmunity. Their long-term use may still be associated with problems arising from non-specific suppression of the immune system. With this in mind, a number of groups have been investigating methods for reinstating antigen-specific tolerance in the adult. One approach which has received much attention recently has been mucosal tolerance. This is based on the observation that antigens encountered either by ingestion or inhalation are generally tolerated by the immune system. A great deal of recent work has demonstrated that oral tolerance to antigens is an active process mediated by antigen-specific lymphocytes. This led to the belief that the administration of those antigens involved in autoimmune disease via mucosal surfaces might reinforce adult tolerance (Thompson & Staines 1990, Weiner et al 1994). Mucosal tolerance has been extensively investigated in rat models of experimental autoimmune encephalomyelitis (EAE) by the groups of Whitacre and Weiner (Miller et al 1991, Whitacre et al 1991). These two groups have defined a number of mechanisms for oral tolerance ranging from activation-induced cell death through functional anergy to modulation of cytokine production among autoantigen-specific T lymphocytes. The mechanism of oral tolerance appears to depend on the dose of antigen administered (Gregerson et al 1993). Oral feeding with low amounts of antigen has been shown to induce CD4 cells producing cytokines such as transforming growth factor (TGF)-β and interleukin (IL)-4 (Chen et al 1994) while administration of higher does of antigen can either lead to functional anergy (Whitacre et al 1991) or apoptosis among potentially autoreactive cells (Chen et al 1995). Since oral myelin is currently being administered to MS patients in phase 3 trials, we chose to study the mechanism of this approach in an alternative mouse model of EAE.

Mucosal tolerance in a mouse model of EAE

Over recent years we have been studying mucosal tolerance in the H-2^u mouse model of EAE. Previously published observations have shown that it is relatively difficult to induce oral tolerance in this model of EAE. High doses of myelin basic protein (MBP) failed to modify the course of disease induced either by peptide antigen or MBP itself (Metzler & Wraith 1996). Furthermore, feeding high doses of whole myelin also failed to modify the course of disease. Our failure to induce oral tolerance in this particular model clearly requires an explanation in light of the success of other groups using the Lewis rat model (Miller et al 1991, Whitacre et al 1991). We reasoned that the lack of induced tolerance in our model might be a consequence of the surprisingly weak antigenic properties of the immunodominant epitope involved in this disease. The immunodominant epitope of MBP maps to the N-terminal, acetylated nonamer. Previous studies have shown that this epitope binds with extraordinarily low affinity to its MHC-restriction element (Fugger et al 1996, Mason et al 1995), thus allowing

TABLE 1 Comparison of oral and nasal antigen delivery in protecting against peptide-induced EAE

Group[a]	*Antigen*	*Route*	*Incidence of disease*	*Median day of onset*	*Mean maximal grade of EAE*
2 × buffer	—	Oral	8/8	12	1.7
2 × 400 μg	Ac1–11[4K]	Oral	6/6	11.5	2.0
2 × 400 μg	Ac1–11[4A]	Oral	6/6	10	2.0
3 × buffer	—	Oral	6/10	15.5	1.2
3 × 1 mg	Ac1–9[4Y]	Oral	0/10	—	0
1 × buffer	—	Nasal	9/10	13	1.2
1 × 100 μg	Ac1–11	Nasal	4/10	> time of experiment	0.5
1 × 100 μg	Ac1–11[4A]	Nasal	2/10	> time of experiment	0.2
1 × 100 μg	Ac1–11[4Y]	Nasal	0/10	—	—

[a]Mice of the H-2^u haplotype either inhaled or were fed the indicated amount of peptide in buffer on days −7, −5 and −3 (3 ×), days −7 and −5 (2 ×) or day −7 (1 ×). On day 0, EAE was induced with the encephalitogenic peptide in CFA (Metzler & Wraith 1993).

T lymphocytes specific for MBP to escape tolerance induction in the thymus (Liu et al 1995). Substitution of the wild-type lysine residue at position 4 of the peptide with more hydrophobic amino acids has been shown to generate analogues with both higher affinity for the class II MHC protein and improved antigenic properties *in vitro* (Fairchild et al 1993). As expected, the low-affinity wild-type encephalogenic epitope of MBP failed to induce oral tolerance over a wide dose range (Metzler & Wraith 1993). An analogue with intermediate affinity also failed to induce oral tolerance when administered at relatively high concentrations. Oral tolerance could be induced, however, by feeding the highest-affinity analogue containing a tyrosine residue at position 4 (Ac1–9(4Y)).

In parallel experiments, while oral administration of the wild-type, encephalitogenic epitope of MBP failed to suppress the induction of EAE, nasal deposition of the same epitope had a significant effect on disease severity in the H-2^u mouse model. As shown in Table 1, even the wild-type peptide itself had an influence on disease severity. Both the intermediate affinity (4A) and the high affinity (4Y) analogues of MBP had an increasingly suppressive effect when administered prior to disease induction. Indeed, in the particular experiment displayed in Table 1, the high affinity analogue of MBP completely prevented induction of disease by the wild-type, encephalitogenic epitope. In addition, nasal peptide treatment effectively reduced relapse rate in treatment of chronic-relapsing EAE (Metzler & Wraith 1996).

Bystander regulation following peptide administration

It is relatively easy to explain how peptides, and in particular highly antigenic peptides, might induce tolerance among antigen-specific T cells. In humans, however, autoimmune diseases are likely to involve T cell responses directed to a variety of antigens. There is a wide range of potential antigens in myelin amongst which the most common are MBP, proteolipid protein (PLP), myelin oligodendrocyte glycoprotein (MOG), myelin-associated glycoprotein and cyclic nucleotidyl phosphodiesterase (CNP). How could a single peptide epitope possibly modify a complex disease involving so many potential auto-antigens and auto-antigenic T cell epitopes? Figure 1 displays some of the known encephalitogenic T cell epitopes in the commonly used H-2^u and H-2^s strains of mice. MBP contains at least three epitopes Ac1–9(A^u), 35–47(E^u) and 89–101(A^s). PLP contains two epitopes, 139–151 and 178–191, both of which are A^s restricted. The (B10.PL × SJL).F1 (H-2^u × H-2^s) mouse is highly susceptible to the induction of EAE with whole myelin and following such immunization responds well to each of the epitopes mentioned above. In order to investigate peptide therapy in this experimental model, we have tested four separate protocols designed to measure either lymphocyte priming or the induction of EAE following immunization with various antigens. In the first protocol (Table 2a), the immunodominant peptide of PLP(139–151) was administered intranasally prior to antigenic challenge with whole myelin in complete Freund's adjuvant (CFA). Draining lymph node cells were subsequently restimulated *in vitro* with peptides Ac1–9, 89–101 or 139–151. Prior intranasal administration of the single epitope from PLP effectively prevented priming of T cells not only to the homologous epitope but also to the two epitopes from MBP. Nasal administration of peptide also both prevented the induction of EAE and treated

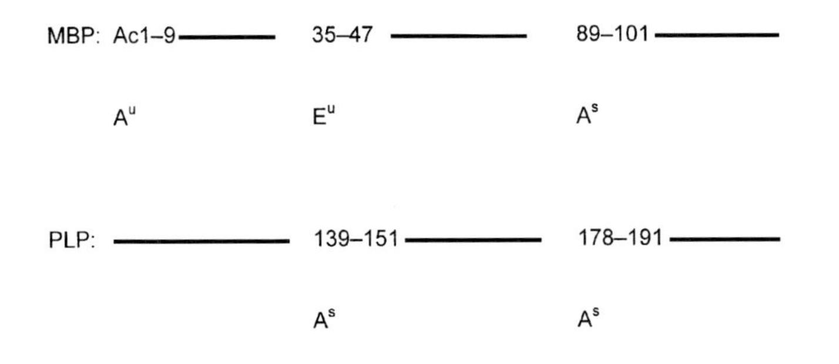

FIG. 1. Encephalitogenic epitopes in mouse myelin. This figure displays the known peptide epitopes in myelin basic protein (MBP) and proteolipid protein (PLP) that are capable of inducing EAE in mice of the H-2^u and H-2^s haplotypes.

TABLE 2 Bystander regulation of the immune response to myelin epitopes following nasal peptide administration

Pre-treat IN[a]	*Challenge SC*	*Challenge in vitro*	*Proliferation*
(a)			
PBS	Myelin	Ac1–9	+++
PBS	Myelin	89–101	+++
PBS	Myelin	139–151	+++
139–151	Myelin	Ac1–9	−
139–151	Myelin	89–101	−
139–151	Myelin	139–151	−
(b)			
PBS	Ac1–9	Ac1–9	+++
PBS	89–101	89–101	+++
PBS	139–151	139–151	+++
139–151	Ac1–9	Ac1–9	+++
139–151	89–101	89–101	+++
139–151	139–151	139–151	−

[a]u × s mice received 100 μg intranasal (IN) doses of peptide or PBS on days −8, −6 and −4 prior to challenge by subcutaneous (SC) injection with myelin in CFA on day 0.
[b]u × s mice received 100 μg doses of peptide or PBS on days −8, −6 and −4 prior to SC priming with the indicated peptides in CFA on day 0. On day 10, draining lymph nodes were disrupted and cultured with the indicated peptide to assess their proliferative capacity by [^{3}H]thymidine incorporation.

ongoing disease, following immunization with whole myelin. These results show that intranasal administration of a single peptide epitope can induce bystander regulation of the immune response to T cell epitopes contained within a complex tissue such as myelin. Next we chose to dissect this phenomenon further by looking for bystander regulation either at the induction or the effector phase of the T lymphocyte response to myelin antigens. Table 2b shows the results of a representative experiment in which lymphocytes from mice pretreated by intranasal deposition of peptide 139–151 failed to respond to the homologous peptide following immunization with the peptide in CFA. Mice primed with isolated heterologous peptides derived from MBP, however, responded perfectly well to these epitopes. In the inductive phase there is, therefore, a high degree of specificity for the inhaled peptide antigen. Regulation of draining lymph node responses to other antigens within myelin requires immunization with the intact antigen since immune responses to isolated epitopes are unaffected by prior administration of the single epitope alone. As shown in Table 3, however, disease induced by immunization with either of the two epitopes from MBP was

TABLE 3 Inhalation of a single peptide suppresses EAE induced with PLP or MBP T cell epitopes

	Mean maximal grade EAE in disease induced with:		
Pre-treatment[a]	*139–151*	*Ac1–9*	*89–101*
PBS	2.6	2.7	3.2
139–151	0.8	1.2	1.5

[a]Mice (nine per group) received three intranasal doses of 100 μg PLP[139–151] or PBS on days −8, −6 and −4 prior to EAE induction with the indicated peptide on day 0. Similar results were obtained in two further experiments.

nevertheless suppressed by prior administration of the epitope from PLP. The difference between the results shown in Tables 2 and 3, following immunization with isolated peptides, may be explained as follows. Nasal administration of the PLP peptide recruits T lymphocytes with specificity for the peptide antigen but with an overall regulatory phenotype (Fig. 2). When the same animal is immunized with a heterologous MBP peptide in CFA (Table 2b), there is no reason why the PLP-specific regulatory cells should target antigen-presenting cells (APCs) presenting the heterologous peptide in the draining lymph node. Hence, PLP-specific regulatory cells fail to prevent priming of MBP-specific T cells when animals are primed with an isolated MBP peptide in CFA. When, however, the animal is immunized with myelin in CFA, these same APCs will present epitopes from both proteins resulting in suppression of the response to the epitope from MBP (Table 2a). Likewise, during the course of disease (EAE) induced by immunization with a single epitope from MBP, one would expect APCs, associated with lesions in or lymph nodes draining the CNS, to present epitopes derived from both protein antigens. Even though disease may have been induced with a single epitope from MBP, it would be expected that regulatory T cells specific for heterologous epitopes from PLP would still suppress disease, as shown in Table 3.

Successful prevention of a number of different experimental autoimmune disease models by intranasal peptide treatment has now been described. Nasal administration of a peptide epitope from heat shock protein 65 has been shown to both prevent and treat arthritis induced by either mycobacterium or avridine in the Lewis rat (Prakken et al 1997). Likewise a single peptide epitope from collagen type II was shown to induce bystander suppression of collagen-induced arthritis in the WA rat (Staines et al 1996). Most notably, however, peptides from two distinct β-islet antigens have been shown to prevent the onset of diabetes in the NOD (non-obese diabetic) mouse. Nasal administration of a peptide from insulin led to reduction in the proliferative response of lymphocytes to the peptide epitope

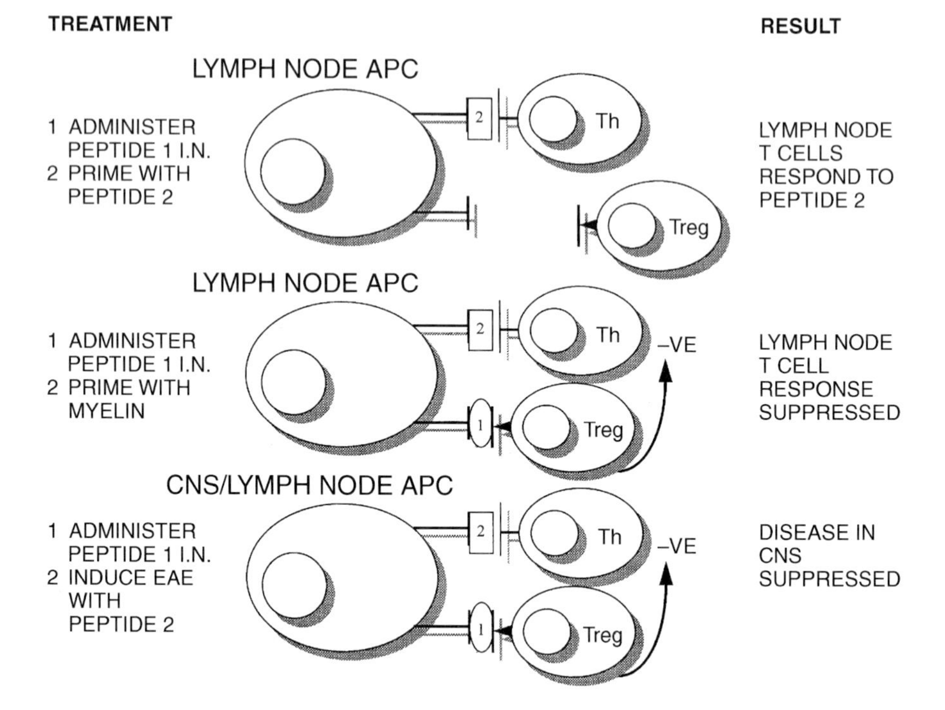

FIG. 2. Bystander regulation of T cell responses in EAE. This figure shows three of the experimental protocols described in the text illustrating the regulatory activity of cells induced by intranasal (IN) administration of peptide. The regulatory T cells (Treg) induced by IN administration of peptide (peptide 1) are capable of suppressing the T helper cell response (Th) to a second T cell epitope (peptide 2) but only when the two epitopes are presented by the same antigen-presenting cell (APC). This can occur either in the draining lymph node following immunization with myelin or in the CNS, or lymph nodes draining the CNS, during the course of disease.

even though these cells retained the capacity to produce T helper 2 (Th2) cytokines (Daniel & Wegmann 1996). Nasal administration of peptides from glutamate decarboxylase (GAD) elicited CD4 cells of a Th2 type which were capable of inhibiting transfer of diabetes in the NOD-SCID mouse model (Tian et al 1996). In this case, bystander suppression of disease and inhibition of determinant spreading in the NOD model of diabetes could both be accounted for by secretion of anti-inflammatory cytokines by Th2 cells.

We have investigated whether or not bystander regulation in the EAE model can also be accounted for by the induction of Th2 cells. This was measured by analysis of both cytokine production from immunized mice *in vitro* and the isotype of anti-MBP antibodies elicited by immunization with MBP *in vivo*. Production of both IFN-γ and IL-4 by antigen-specific $CD4^+$ T cells was inhibited by prior intranasal administration of peptide antigen. Furthermore, the induction of antibodies of both IgG1 and IgG2a isotypes was suppressed by prior intranasal deposition of peptide antigen. This clearly argues against a shift from Th1 to Th2 lymphocyte responses in the bystander regulation of EAE in this particular model. What then is the mechanism?

Peripheral tolerance in a transgenic model of EAE

The effect of soluble peptide administration is most readily revealed in mice with a high frequency of the relevant T cell receptor. Mice expressing a T cell receptor specific for the N-terminal peptide of MBP (Liu et al 1995) were treated intranasally with peptide analogues of varying affinity. This led to transient activation of the majority of cells in both the draining and distant lymphoid organs. Activation was revealed by up-regulation of both CD69 and the IL-2 receptor, with concomitant down-regulation of the T cell receptor as had previously been shown to follow intraperitoneal treatment with soluble peptide (Liu et al 1995). Cell surface marker expression returned to normal levels within 24 hours. Nasal administration of peptide antigen led to apoptosis among CD4/8 double-positive cells in the thymus and the appearance of some hypodiploid CD4 single-positive cells in peripheral lymphoid organs, indicative of apoptosis in mature lymphocytes. A significant proportion of cells, however, survived the intranasal administration of soluble peptide antigen. Among these cells a proportion underwent cell division as revealed by the staining pattern of lymphocytes labelled with CFSE (5[6]-carboxyfluorescein diacetate succinimidyl-ester), a dye used to trace dividing cells *in vivo* (Lyons & Parish 1994).

The most striking effect of nasal peptide administration was revealed by analysis of cytokine production. In parallel with the transient hyperactivation revealed by cell surface marker staining, cells tested directly from mice treated with intranasal peptide produced a transient burst of cytokines representative of both Th1 and Th2 cells (Table 4). The hyperactivation of cytokine production was, however, prevented by a single prior intranasal administration of peptide. As well as causing transient activation and apoptosis among antigen-specific CD4 cells, the encounter with soluble peptide antigen appeared to lock the remaining transgenic cells into a state where they were no longer capable of producing cytokines such as IL-2, IL-4 and IFN-γ. Despite the high frequency of cells expressing a single T cell receptor in the transgenic mouse, these could, nevertheless, be rendered unresponsive by repeated nasal administration (Table 5). Unresponsiveness was particularly marked in assays designed to measure either proliferation or Th1 cytokine production. The wholesale down-regulation of the immune response to antigen following intranasal peptide administration did not, however, extend to IL-10. As seen in Table 5, the production of IL-10 was surprisingly resistant to

TABLE 4 Prevention of T cell hyperactivation by intranasal peptide administration

Mice pre-treated IN with: (day)	*Ex-vivo cytokine production*[a]		
	IFN-γ	*IL-2*	*IL-4*
PBS (d − 1)	–	–	–
Ac1–9[4Y] (d − 1)	+++	+++	+
Ac1–9[4Y] (d − 8+d − 1)	–	–	–

[a]Transgenic mice expressing the T cell receptor specific for peptide Ac1–9 of MBP were treated with intranasal (IN) peptide or PBS on day −1 or days −8 and −1. Cervical lymph node cells were disrupted and assessed directly for cytokine production using the cell-based ELISA method described by Beech et al (1997).

TABLE 5 Selective induction of tolerance in TCR-transgenic T cells following nasal peptide administration

Pre-treatment[a]	*Response to antigenic challenge in vitro*				
	Proliferation	*IFN-γ*	*IL-2*	*IL-4*	*IL-10*
PBS IN	+	++	++	–	–
[4Y] 1 × IN	+++	++	++	++	++
[4Y] 5 × IN	–	–	–	–	++
[4Y] 10 × IN	–	–	–	–	++

[a]Transgenic mice expressing the T cell receptor (TCR) specific for peptide Ac1–9 of MBP were treated with intranasal (IN) peptide (Ac1–9[4Y]) or PBS prior to antigenic challenge with Ac1–9 *in vitro*. Mice either received a single dose of peptide 24 h or alternatively 5 or 10 doses of peptide every other day before challenge *in vitro*. Cervical lymph node cells were disrupted, challenged *in vitro* with Ac1–9 and assessed either for proliferative capacity by [^{3}H]thymidine incorporation (Liu et al 1995) or cytokine production using the cell-based ELISA method described by Beech et al (1997).

repeated antigen administration. Naïve cells, therefore, respond to encounter with soluble peptide by transient activation after which they become resistant to further activation, both in terms of selected cytokine production and proliferation and yet they retain the capacity to secrete IL-10.

The results of this study may be summarized as follows:

(1) Intranasal administration of peptide can effectively prevent and treat EAE.
(2) There is clear evidence of bystander regulation following intranasal peptide administration.
(3) There is a direct relationship between antigenicity and tolerogenicity.
(4) There is no evidence for a wholesale switch from Th1 to Th2 responses.

The results described here clearly distinguish the EAE model from the NOD model of intranasal peptide tolerance. Why should intranasal peptide induce Th2 cytokines in one model while suppressing the same cytokines in another? It seems unlikely that this is specific for the peptide antigen since a variety of peptides have been shown to induce Th2 responses in the NOD mouse while peptides from both PLP and MBP were shown to inhibit Th2 cytokines in the EAE model. There are two other possibilities. First, in order to prevent disease in the NOD model, mice were treated with peptide from a very young age. It could be that treatment of very young animals with soluble peptide tends to favour Th2 responses, as suggested by recent studies of neonatal tolerance (Forsthuber et al 1996). Secondly, there could be an inherent genetic difference between the mice used in our EAE studies and the NOD mouse. It will be crucial to clarify this issue in future experiments. Recent studies have shown that Th2 cells may not always be protective against diseases induced by Th1 cells. Genain et al (1996) have shown that the induction of Th2 cells in a marmoset model of EAE can exacerbate disease by increasing the level of pathogenic antibodies. This result demonstrates the potential danger of antigen-specific therapy. It is too simplistic to believe that induction of Th2 responses will protect individuals against diseases commonly associated with inflammatory (Th1) responses. This latter result emphasizes how much we have to learn about the mechanism of antigen-specific tolerance before this approach can safely be applied for routine treatment of autoimmune diseases in the clinic.

The results discussed in this manuscript reveal a wide variety of responses to intranasal peptide administration. These range from apoptosis through proliferation to anergy induction. In our transgenic model, repeated administration of peptide antigen results in apparent anergy as revealed by suppressed proliferation and cytokine production. One clear exception to this pattern is IL-10. Despite the lack of proliferation and inhibition of selected cytokines, including IL-2, IL-4 and IFN-γ, cells subjected to repeated encounter with peptide antigen retain the capacity to produce IL-10. This observation confirms previous studies in a mouse model of superantigen-induced anergy (Sundstedt et al 1997). In this model, mice expressing a single Vβ3 T cell receptor were subjected to repeated administration of the superantigen staphylococcal enterotoxin A (SEA). A single administration of SEA resulted in hyperactivation and cytokine production. Repeated administration, however, inhibited the production of most cytokines but had little effect on IL-10 production. The question remains as to whether production of IL-10 correlates with the phenomenon of bystander regulation. Interestingly, IL-10 has also been shown to induce a long-term antigen-specific anergic state in human CD4$^+$ T lymphocytes *in vitro* (Groux et al 1996). Confirmation that this cytokine is responsible for bystander regulation *in vivo*, however, awaits further experimentation.

Acknowledgements

The authors are indebted to the Wellcome Trust and the Multiple Sclerosis Society of Great Britain and Northern Ireland for financial support. C. Burkhart acknowledges receipt of a Research Fellowship from the Deutsche Forschungsgemeinschaft.

References

Beech JT, Bainbridge T, Thompson SJ 1997 Incorporation of cells into an ELISA system enhances antigen-driven lymphokine detection. J Immunol Methods 205:163–168

Chen Y, Kuchroo VK, Inobe J-I, Hafler DA, Weiner HL 1994 Regulatory T cell clones induced by oral tolerance: suppression of autoimmune encephalomyelitis. Science 265:1237–1240

Chen Y, Inobe JI, Marks R, Gonnella P, Kuchroo VK, Weiner HL 1995 Peripheral deletion of antigen-reactive T cells in oral tolerance. Nature 376:177–180

Daniel D, Wegmann DR 1996 Protection of nonobese diabetic mice from diabetes by intranasal or subcutaneous administration of insulin peptide B-(9–23). Proc Natl Acad Sci USA 93:956–960

Fairchild PJ, Wildgoose R, Atherton E, Webb S, Wraith DC 1993 An autoantigenic T cell epitope forms unstable complexes with class II MHC: a novel route for escape from tolerance induction. Int Immunol 5:1151–1158

Forsthuber T, Yip HC, Lehmann PV 1996 Induction of Th1 and Th2 immunity in neonatal mice. Science 271:1728–1730

Fugger L, Liang J, Gautam A, Rothbard JB, McDevitt HO 1996 Quantitative analysis of peptides from myelin basic protein binding to the MHC class II protein, I-A^u, which confers susceptibility to experimental allergic encephalomyelitis. Mol Med 2:181–188

Genain CP, Abel K, Belmar N et al 1996 Late complications of immune deviation therapy in nonhuman primate. Science 274:2054–2057

Gregerson DS, Obritsch WF, Donoso LA 1993 Oral tolerance in experimental autoimmune uveoretinitis. Distinct mechanisms are induced by low dose vs high dose feeding protocols. J Immunol 151:5751–5761

Groux H, Bigler M, de Vries JE, Roncarolo M-G 1996 Interleukin-10 induces a long-term antigen-specific anergic state in human $CD4^+$ T cells. J Exp Med 184:19–29

Liu GY, Fairchild PJ, Smith RM, Prowle JR, Kioussis D, Wraith DC 1995 Low avidity recognition of self-antigen by T cells permits escape from central tolerance. Immunity 3:407–415

Lyons AB, Parish CR 1994 Determination of lymphocyte division by flow cytometry. J Immunol Methods 171:131–137

Mason K, Denney DW, McConnell HM 1995 Myelin basic peptide complexes with class II MHC molecules I-A^u and I-A^k form and dissociate rapidly at neutral pH. J Immunol 154:5216–5227

Metzler B, Wraith DC 1993 Inhibition of experimental autoimmune encephalomyelitis by inhalation but not oral administration of the encephalitogenic peptide: influence of MHC binding affinity. Int Immunol 5:1159–1165

Metzler B, Wraith DC 1996 Mucosal tolerance in a murine model of experimental autoimmune encephalomyelitis. Ann N Y Acad Sci 778:228–243

Miller A, Lider O, Weiner HL 1991 Antigen-driven bystander suppression after oral administration of antigens. J Exp Med 174:791–798

Prakken BJ, van der Zee R, Anderton SM, van Kooten P, Kuis W, van Eden W 1997 Peptide induced nasal tolerance for a mycobacterial hsp60 T cell epitope in rats suppresses both adjuvant arthritis and non-microbially induced experimental arthritis. Proc Natl Acad Sci USA 94:3284–3289

Staines NA, Harper N, Ward FJ, Malmstrom V, Holmdahl R, Bansal S 1996 Mucosal tolerance and suppression of collagen-induced arthritis induced by nasal inhalation of synthetic peptide 184–198 of bovine type II collagen expressing a dominant T cell epitope. Clin Exp Immunol 103:368–375

Sundstedt A, Hoiden I, Rosendahl A, Kalland T, van Rooijen N, Dohlsten M 1997 Immunoregulatory role of IL-10 during superantigen-induced hyporesponsiveness *in vivo*. J Immunol 158:180–186

Thompson HSG, Staines NA 1990 Could specific oral tolerance be a therapy for autoimmune disease? Immunol Today 11:369–399

Tian J, Atkinson MA, Clare-Salzer M et al 1996 Nasal administration of glutamate decarboxylase (GAD 65) peptides induces Th2 responses and prevents murine insulin-dependent diabetes. J Exp Med 183:1561–1567

Weiner HL, Friedman A, Miller A et al 1994 Oral tolerance: immunologic mechanisms and treatment of animal and human organ-specific autoimmune diseases by oral administration of autoantigens. Annu Rev Immunol 12:809–838

Whitacre CC, Gienapp IE, Orosz CG, Bitar DM 1991 Oral tolerance in experimental autoimmune encephalomyelitis. III. Evidence for clonal anergy. J Immunol 147:2155–2163

Wraith DC 1998 Current and future therapies for Multiple Sclerosis. In: Scolding NJ (ed) Immunological and inflammatory diseases of the central nervous system. Butterworth-Heinemann, Oxford, in press

DISCUSSION

Hafler: One of the key points to emerge from our work is that the dose of antigen that one administers is critical in determining whether one generates either IL-4- or TGF-β-secreting cells. TGF-β secretion depends upon the particular model and the antigen. In the NOD mouse one doesn't see TGF-β; one sees more IL-4, for example. But the key factor appears to be dose of antigen. With low doses one generates these Th2- or Th3-type cells; with high doses we see exactly what you see, which is induction of apoptosis or anergy. The question is, have you reduced the dose of antigen you have administered to the extent that one might be able to see these different responses?

Wraith: The original experiments that Barbara Metzler did were done over a nanogram to milligram range of dose response. With the wild-type MBP peptide she never saw any induction of oral tolerance, which differs markedly from what you found with PLP peptide in SJL mice.

Hafler: Or the MBP peptide. Did you do it in SJL mice?

Wraith: It was done in the H-2^u and the H-2^u × H-2^s.F1, but only with the MBP epitope. I acknowledge that the more highly antigenic epitopes, such as 139–151, can be tolerogenic by oral administration. I was trying to emphasize strongly that intranasal administration appears to be better than oral.

Hafler: The data generated by Howard Weiner would suggest that intranasal is as good if not better than oral administration. The critical question is whether it is also operating by this mechanism of bystander suppression, or with intranasal administration is one seeing what in effect is a high intravenous dose using the mucosal system as a filter?

Wraith: This possibility cannot be excluded.

Mitchison: A point of clarification: was your IL-10 experiment done with transgenic T cells?

Wraith: Yes. If you do the same experiment with the normal mouse, you can't detect IL-10.

Mitchison: Is that because there is not enough of it?

Wraith: I assume so.

Hafler: Without blocking with anti-IL-10, you don't really know whether it is important or not.

Shevach: You cautiously avoided using the word 'therapeutic' in your talk: could you comment on this?

Wraith: I did mention therapy of the NOD mouse, and we have also treated chronic-relapsing EAE models after disease onset. Basically, if you have a chronic-relapsing disease and you treat immediately after disease onset then the treatment may not have an immediate effect but it will affect subsequent relapses. Barbara Metzler has published most of this work (Metzler & Wraith 1996). Sometimes you see exacerbation of disease immediately after treatment, sometimes you see down-regulation in that first round, but you always see a dampening down of the relapses. This fits very well with the fact that it takes about four days for the tolerance effect to be seen after priming (Liu & Wraith 1995).

Mason: In principle, bystander suppression could be regarded as rather a dangerous phenomenon, since you might suppress responses you don't want to suppress. If you actually look at responses to a real foreign antigen, as opposed to a self-antigen for which the T cell affinity may already be modulated by negative selection, do you see a difference?

Wraith: In all the experiments that I've shown you where we've primed with the peptide in CFA after nasal tolerization with the same peptide, we wipe out the response to the peptide, but we have no effect whatsoever on the response to purified protein derivative (PPD). As you know, PPD is a control antigen contained in the mycobacteria included in CFA.

Hafler: The clinical reality is that we now often treat patients with 1200 mg/month of cyclophosphamide, so relatively speaking there is always going to be a price to pay for any therapy.

Mason: Yes, but this is an important physiological distinction between a self antigen and a non-self antigen.

Abbas: I want to quibble with the use of data from intact TCR transgenics to conclude that apoptosis is a mechanism by which oral antigen works. We've known for aeons that it doesn't matter where you put the antigen, you're going to get death in local draining lymphoid tissues. It has got nothing to do with oral antigen, it's just the result of putting a lot of antigen into a mouse with 30% of its T cells specific for the antigen.

Wraith: I thought I was saying that apoptosis wasn't responsible for the peripheral tolerance that we observe.

Abbas: It has been published that it is responsible and that's a TCR transgenic experiment.

Wraith: I was trying to emphasize that you can detect apoptosis.

Abbas: But you don't think that is the mechanism?

Wraith: Absolutely not. How can you explain bystander suppression through apoptosis?

Mitchison: I think you lost everybody when you said 30% of the cells were undergoing apoptosis in your controls: isn't that rather incredible?

Lenardo: There is a problem with that interpretation. Many things cause phosphatidyl serine display, including platelet aggregation, so using annexin as an indication of cell death may not be completely accurate if no other parameters are investigated.

Wraith: Can we modify it by saying that the background of annexin-positive cells is 30%?

Mason: The rate of apoptosis must equal the rate at which you're generating new cells by cell division, since the animal doesn't keep filling up with cells. That is therefore the natural rate of apoptosis regardless of the mechanism.

Wraith: There has to be a natural turnover of cells, but what these guys can't stomach is the fact that it is 30%.

Lenardo: In my view, bystander suppression is an anathema, because to pursue such an effect, one has lost the specificity of antigen-induced immunosuppression. If you put in some antigen hoping to suppress other T cells that react to a completely different antigen to get disease suppression, I think it is unlikely to work and disregards the exquisite antigen specificity in the immune system. For example, in the paper by Weiner and co-workers (Miller et al 1991), bystander suppression was not even dealing with neural antigens and used antigens such as ovalbumin and bovine serum albumin to obtain suppression *in vitro* that was modest.

Hafler: That's not true. In our paper published in *Science*, oral administration of PLP suppressed MBP-induced disease (Chen et al 1994).

Lenardo: But how many other things were compared in that paper? Did you look at the effects of the few clones that you described on a variety of different antigens to see if there was a global suppression? So my point is this: if we are talking about a fairly minor effect of 50% suppression of a proliferation *in vitro*, which is what was described in the Weiner paper (Miller et al 1991), I don't think that this is strong evidence for a clinically useful effect. The fact that the clinical trials based on this theory have failed further supports my point.

Wraith: Michael, I take strong exception to your statement that we are seeing a fairly minor, 50% effect. I show that the proliferation is reduced from 25 000 cpm to less than 500 cpm, which is the equivalent of background. These are the data: you can't argue against the data.

Lenardo: I disagree strongly with the idea that bystander suppression is necessary because of the complexity of the antigens that are potentially involved in autoimmune diseases. My belief is that antigen-specific treatments to suppress T cell-mediated diseases will become the method of choice.

Allen: A simple point: you are using the NOD mouse as the example for bystander suppression. Does it have to be the same APC presenting all these epitopes to be able to see this?

Wraith: I showed the NOD mouse data to demonstrate that some people have clearly demonstrated a Th1–Th2 shift. We need to think about why the NOD mouse responds in that way and why we don't see the same thing. It could be that you're treating the NOD mouse from a very young age. It might be this that is setting the Th2 phenotype. It could also be that the NOD mouse is so genetically different from other mice. I wasn't using that as an example of all the antigens being presented by the same APCs, but in a way I wouldn't be surprised if that were so.

Hafler: Have you looked in mesenteric lymph nodes?

Wraith: We've looked. Most of that was for survival.

Hafler: A lot of the work done on oral tolerance relates to mucosal lymph nodes, which are different.

Mitchison: If you look more widely, at other negative regulatory cells, it's clear that in some parasite systems the protective or suppressive cytokine is IL-4, and in other systems it is TGF-β. It's very nice to have IL-10 join the throng, but it is becoming crowded now, isn't it?

Wraith: Yes, but we've looked hard for TGF-β and IL-4, and if we had been able to find those we might have been able to publish these data two years ago.

Shevach: Just one comment in support of IL-10. We have recently looked at a panel of knockout C57BL/10 mice and immunized them with MBP. Although wild-type C57BL/10 have a low (20%) incidence of EAE, the C57BL/10 IL-10 knockout mice are much more susceptible to EAE, with 50–60% of the mice getting quite sick.

Abbas: But in the C57BL/10 IL-10 knockout the macrophages are so out of control that it is just as susceptible to turpentine painting of the skin. That experiment doesn't necessarily implicate IL-10 as a physiological down-regulator of EAE.

Shevach: We have other studies which suggest that the cell responsible for IL-10 production is actually a $CD4^+$ T cell.

Waldmann: What are the consequences of immunizing the animal in a bystander-suppressed system? When you give PLP and you don't get disease, what's the animal's long-term record of encounter with PLP?

Wraith: In other words, if we come back three months later and immunize, what happens? Is the animal tolerant?

Barbara Metzler has done experiments in normal and thymectomized animals. She treated them with soluble peptide nasally and then waited one, two, four or eight weeks before priming with antigen in CFA. The animals are tolerant at least out to eight weeks.

Waldmann: So the T cell that is initially regulatory comes into the environment of the second T cell that has been damped or suppressed. That T cell sees antigen, and as a result may become regulatory itself, but you don't know yet.

Hafler: In our model tolerance lasts about three months.

Jenkins: If I understood you, David, the reason we did our experiment incorrectly was that we didn't link the antigens. In terms of what we understand about antigen processing, what does that mean?

Wraith: If, let's say, IL-10 is responsible for this effect, then it is quite simple: IL-10 is preventing presentation of the linked epitope.

Jenkins: But when you mix the antigens together, the same APCs are going to present both antigens.

Hafler: I disagree: I don't think the issue is linking. The issue is inducing T cells that have a different cytokine profile. You don't have to have antigen linkage, the antigens have to be in the same organ site.

Wraith: It looks as though it helps.

Shevach: Co-expression of the antigens in the same organ (the brain) is not actually the same as the presentation of linked epitopes.

Hafler: But they're all presented by APCs in the same site. The issue would be to take one set of T cells, grow them up in IL-4 for example, so they are secreting IL-4 and the have the antigen at the same site. Bystander suppression, the phenomenon you are looking at, is a very different one: T cells are entering the organ.

Waldmann: If you had this discussion 30 years ago when people were talking about T–B cooperation, you would have been saying that T cells recognizing the carrier would have been coming into the vicinity of cells recognizing the hapten. There is something about linked recognition that does seem to be important.

Hafler: That is in T–B interaction, but this is a different system.

Mitchison: I don't think anyone has carefully compared linked and unlinked epitopes in suppression, titrating cell numbers and antigen concentration.

Wraith: Not titrating, but equimolar concentrations of two separate peptides injected into a mouse at the same dose as injected whole protein will not give you bystander suppression.

Hafler: But you can get good bystander suppression by transfecting LCMV into the pancreas and feeding LCMV or insulin.

Arnold: Is bystander suppression long-lasting? If so, does it need the presence of the antigen, and is then an anergic T cell *in vivo* a memory-type T cell which when it sees antigen just does not produce IL-2 but something else such as IL-10?

Wraith: The experiment hasn't been done carefully enough.

Arnold: What is an anergic T cell *in vivo* then?

Wraith: I would add a big question mark after anergy. We've been able to demonstrate various elements of unresponsiveness — lack of proliferation, lack of IL-2 production, lack of IFN-γ and lack of IL-4 production — but those same cells appear to be making another cytokine (IL-10), so they're still functional in one way.

Arnold: I'm asking because there are several systems in which it has been shown that the presence of antigen is needed to maintain tolerance.

Wraith: With self-antigens we don't have a problem.

References

Chen Y, Kuchroo V, Inobe J-I, Hafler DA, Weiner HL 1994 Regulatory T cell clones induced by oral tolerance: suppression of autoimmune encephalomyelitis. Science 265:1237–1240

Liu GY, Wraith DC 1995 Affinity for class II MHC determines the extent to which soluble peptides tolerize autoreactive T cells in naive and primed adult mice — implications for autoimmunity. Int Immunol 7:1255–1263

Metzler B, Wraith DC 1996 Mucosal tolerance in a murine model of experimental autoimmune encephalomyelitis. Ann N Y Acad Sci 778:228–242

Miller A, Lider O, Weiner HL 1991 Antigen-driven bystander suppression after oral administration of antigens. J Exp Med 174:791–798

Quantitative and qualitative control of antigen receptor signalling in tolerant B lymphocytes

James I. Healy*, Ricardo E. Dolmetsch†, Richard S. Lewis† and Christopher C. Goodnow‡

**Department of Microbiology and Immunology, †Department of Molecular and Cellular Biology, Stanford University School of Medicine, Stanford, CA 94301, USA and ‡Medical Genome Centre, John Curtin School of Medical Research, Australian National University, Canberra, Australia*

Abstract. Lymphocyte antigen receptors, such as the B cell antigen receptor (BCR), have the ability to promote or inhibit immune responses. This functional plasticity is exemplified by BCR-induced mitosis in naïve but not tolerant B cells and is correlated with biochemical differences in the signals triggered by foreign and self antigens. Acute stimulation of naïve B cells with foreign antigen induces a biphasic Ca^{2+} flux, and activates nuclear signalling through NF-AT, NF-κB, JNK and ERK. In tolerant B lymphocytes, by contrast, self antigen triggers only a low Ca^{2+} plateau, NF-AT and ERK. After removal from self antigen, the BCRs on tolerant B cells reacquire the ability to stimulate a biphasic Ca^{2+} flux and to promote proliferation. The differences in nuclear signalling between naïve and tolerant cells is brought about in part by differences in the magnitude of the Ca^{2+} signal. A low, sustained Ca^{2+} signal, such as that seen in tolerant B cells, activates NF-AT, whereas a high but transient Ca^{2+} spike, which resembles that triggered in naïve B cells, activates NF-κB and JNK. These findings demonstrate that the quantitative differences in Ca^{2+} signalling between naïve and tolerant B cells are reversible and contribute to the differential triggering of nuclear signals. The activation of selected transcription factors may in turn account for the different functional responses triggered in naïve and tolerant lymphocytes.

1998 Immunological tolerance. Wiley, Chichester (Novartis Foundation Symposium 215) p 137–145

Two important features of tolerant lymphocytes indicate that self antigens continually transmit signals to autoreactive cells that are different from those signals induced by foreign antigens in naïve B cells.

Self antigens induce selective responses

The first feature is that ligation of the antigen receptor on tolerant B lymphocytes induces only a subset of the functional responses induced by antigen receptors on

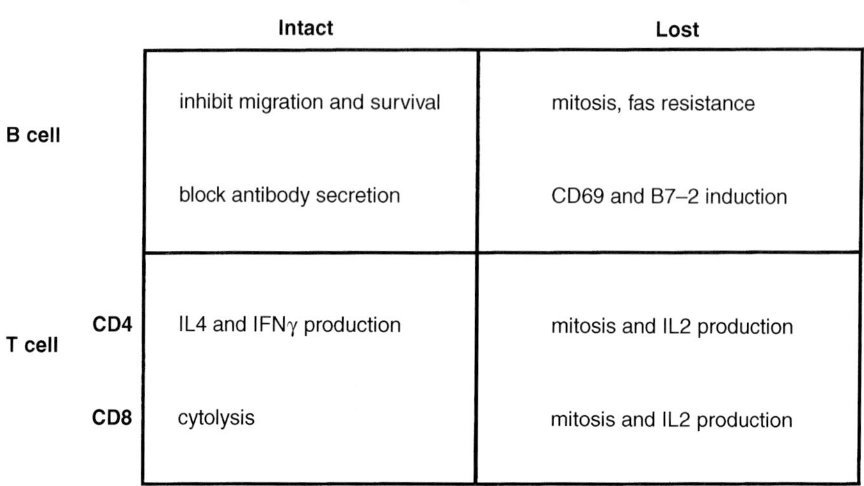

		Responses: Intact	Responses: Lost
B cell		inhibit migration and survival	mitosis, fas resistance
		block antibody secretion	CD69 and B7–2 induction
T cell	CD4	IL4 and IFNγ production	mitosis and IL2 production
	CD8	cytolysis	mitosis and IL2 production

FIG. 1. Table comparing the functional responses which are selectively retained (intact) versus those which are lost in models of B and T cell tolerance.

naïve B cells. In tolerant B cells, for example, self antigen actively inhibits migration (Cyster et al 1994, Cyster & Goodnow 1995), survival (Fulcher et al 1994) and autoantibody secretion (Goodnow et al 1991), but does not trigger mitogenesis (Goodnow et al 1991, Cooke et al 1994), Fas resistance (Rathmell et al 1995) or B7-2 expression (Cooke et al 1994) (Fig. 1, B cells). In anergic T cells, T cell receptor stimulation triggers interleukin (IL)-4 and γ-interferon production in CD4$^+$ cells (Jenkins et al 1987, Mueller et al 1991, Gajewski et al 1994) and cytolysis in CD8$^+$ positive cells (Otten et al 1991, Hollsberg et al 1995). These CD4$^+$ or CD8$^+$ cells are unable to produce normal levels of IL-2 or proliferate in response to antigen receptor ligation (Fig. 1, T cells). The ability of self antigens to trigger only selected functional responses in tolerant B and T cells suggests that autoantigens must induce signals that are different from self antigens.

Self antigens continuously reinforce tolerance

The second feature of tolerant lymphoyctes which suggests that self antigens actively transmit signals is that the maintenance of self-tolerance requires persistent exposure to self antigen. In the absence of antigen the tolerant phenotype is reversible, particularly in the presence of antigen receptor-independent co-stimuli. In immature B cells, for example, removal from multivalent, membrane-bound antigen reverses the developmental arrest which inhibits their maturation (Hartley et al 1993). Likewise, mature B cells that have been rendered tolerant by chronic exposure to soluble self antigen regain surface IgM expression, and can be reactivated to generate antibodies in T cell-dependent or T cell-independent immune responses when they are removed from autoantigen (Goodnow et al 1991). This recovery is only partial in resting cells, but is complete and rapid if tolerant B cells are stimulated into the cell cycle via B cell receptor-independent co-stimuli such as lipopolysaccharide (LPS) or CD40.

The blunted mitogenic response in tolerant B cells, like the inhibition of antibody production, depends on constant self antigen exposure. Transfer of tolerant B cells from hen egg lysozyme (HEL)-expressing to non-HEL expressing mice results in a fourfold higher B cell receptor (BCR)-dependent proliferative response (Fig. 2A, tolerant). This recovery of mitogenic responses in tolerant B cells after removal from self antigen is only partial as naïve cells parked in mice that do not express HEL have at least a twofold higher proliferative response than tolerant B cells that are removed from HEL (Fig. 2A, compare naïve Non with tolerant). Furthermore, after 36 h of antigen exposure, naïve B cells exhibit blunted mitogenic responses (Fig. 2A, compare naïve Non with HEL). Thus inhibition of mitogenic responses is induced by and requires persistent antigen exposure. In the absence of constant antigen receptor engagement the tolerant B cell phenotype is at least partially reversible.

Maintenance of T cell tolerance also depends on continuous antigen exposure. For example, T cell clones that have been rendered anergic only persist in a hyporeactive state for approximately one week (Lamb et al 1983, Jenkins & Schwartz 1987). Similarly, primary T cells which are tolerant to the MHC class I molecule K^b, regain the ability to reject K^b-expressing skin grafts 0.5–2 weeks after transfer into mice that do not express K^b (Alferink et al 1995). The reversible nature of both clonal elimination of B cells and clonal inactivation of B and T cells (Fig. 2B) suggests that constant antigen exposure must induce a signal which reinforces self tolerance but is qualitatively or quantitatively different from the signal induced by acute foreign antigen exposure.

Qualitative differences in signalling

To determine the nature of the signals induced by autoantigens that could maintain the reversible tolerant phenotype, we compared signalling in naïve and tolerant B lymphocytes. Acute foreign antigen exposure to naïve HEL-specific B cells induces a biphasic Ca^{2+} response (Fig. 3, dashed line) and activates signals through NF-AT, NF-κB, JNK and ERK (Healy et al 1997). By comparison, foreign antigen activates a low Ca^{2+} plateau (Fig. 3, bold line), NF-AT and ERK, but not NF-κB or JNK (Healy et al 1997). The transfer of tolerant B cells into mice that do not express HEL reduces the basal Ca^{2+} signalling (Healy et al 1997), and restores the ability to generate a biphasic Ca^{2+} flux in naïve B cells (Fig. 3, dotted

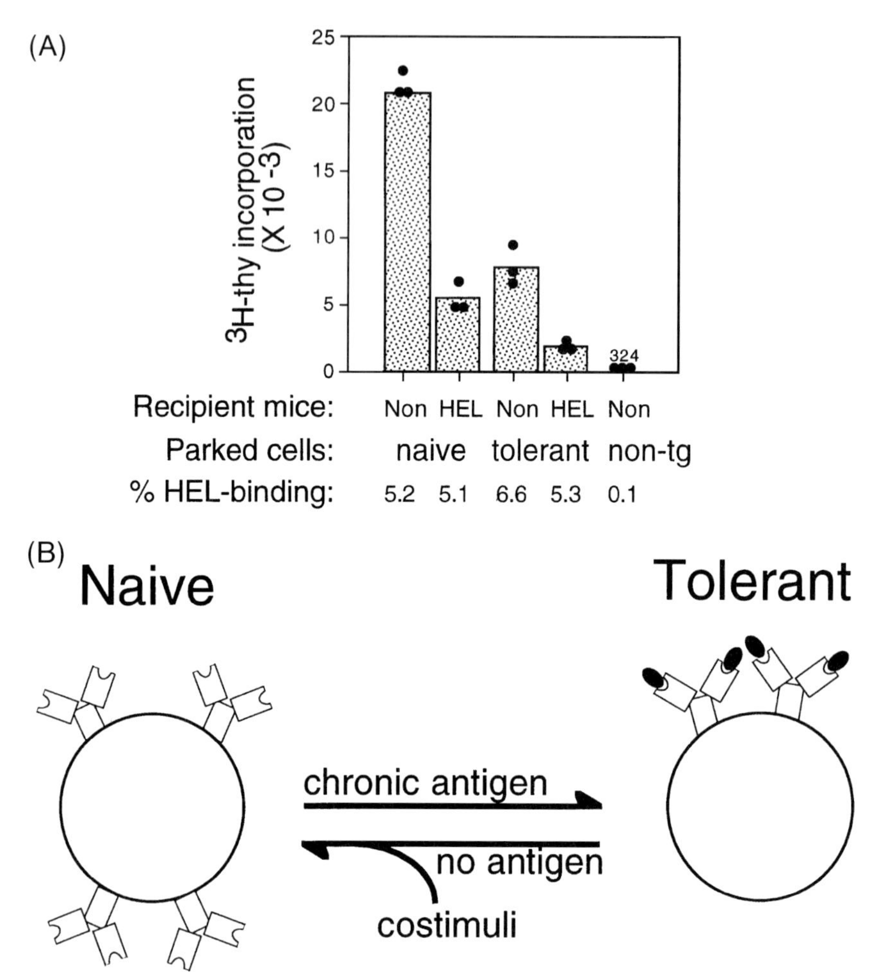

FIG. 2. Reversible B cell tolerance is maintained by persistent antigen exposure. (A) Partial recovery of BCR-induced proliferative response in tolerant cells parked for 36 h in mice lacking HEL. Pairs of sublethally irradiated recipient mice with circulating HEL or lacking HEL (Non) were injected intravenously with 25 million spleen cells from naïve (Ig-HEL) mice, tolerant mice (Ig-HEL/HEL) or non-transgenic (non-tg) cells. Thirty-six hours later, spleen cells from the two recipient mice in each group were purified over Ficoll, the frequency of HEL-binding B cells determined by flow cytometry, and the cells were cultured with anti-IgDa–dextran for 32 h. The cultures were pulsed with tritiated thymidine for the last 12 h in culture and incorporated thymidine was measured form triplicate cultures (dots). Bars show arithmetic means. In other experiments tolerant cells parked in the absence of HEL for 72 h showed no further recovery of the proliferative response. The experiments were performed with E. Adams, whom C.C.G. would like to thank. (B) Model illustrating the reversible nature of the tolerant phenotype. Persistent antigen exposure reinforces tolerance whereas removal from antigen or exposure to co-stimuli such as LPS in the absence of antigen allow tolerant cells to regain a naïve phenotype.

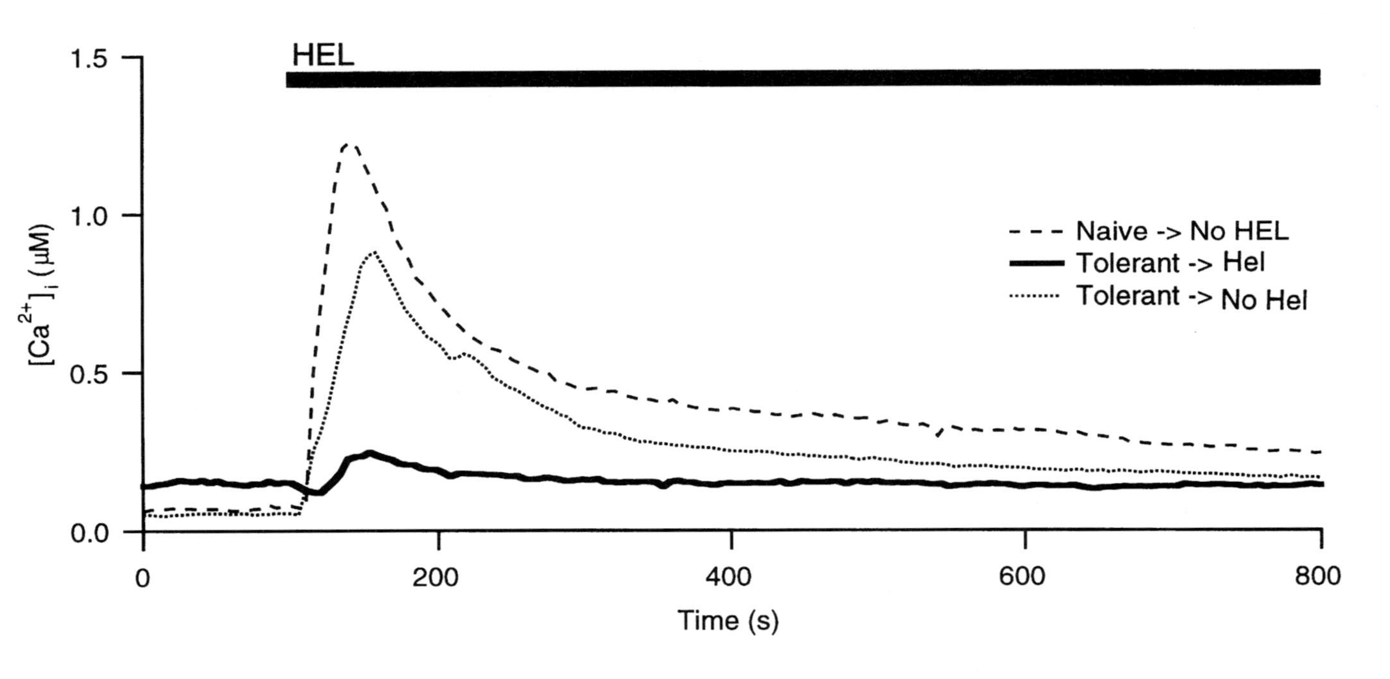

FIG. 3. Partial recovery of biphasic calcium signalling in tolerant B cells 50 h after transfer to mice lacking HEL. Naïve or tolerant B cells were injected intravenously into mice that either did (HEL) or did not (No Hel) express lysozyme. Fifty hours later the cells were purified from the spleens of the recipient mice and analysed by single-cell Ca^{2+} analysis as described in Healy et al (1997). At 100 s the cells were stimulated with 500 ng/ml HEL. Naïve B cells parked in non-HEL-expressing mice are indicated by the dashed line. Tolerant B cells parked in HEL-expressing mice are indicated by the bold line and tolerant B cells transferred to non-HEL-expressing mice are indicated by the dotted line. After recording the single-cell Ca^{2+} imaging data the cells were stained with IgDa–FITC to confirm they were the transgene-expressing transferred B lymphocytes. Similar results were obtained in a separate experiment.

line). The blunted biphasic Ca^{2+} responses in tolerant B cells, like the blunted mitogenic response, is reversible. These findings reveal the remarkable plasticity of signalling by lymphocyte antigen receptors but raise the question of how a single receptor type selectively activates distinct signals such as NF-AT, NF-κB and JNK, which all depend on the second messenger Ca^{2+}.

Quantitative controls

The differential regulation of the two phases of the Ca^{2+} flux and the selective activation of certain transcription factors in tolerant B cells, suggested that the magnitude and duration of the Ca^{2+} response contributes to selective triggering of downstream pathways such as NF-AT, NF-κB or JNK. Naïve B cells exposed to an isolated transient Ca^{2+} pulse exhibited sustained NF-κB and JNK activation but only temporary NF-AT activation. In contrast, a high-level sustained Ca^{2+} signal results in persistent NF-AT, NF-κB and JNK activation (Dolmetsch et al 1997). This demonstrates that transient versus sustained Ca^{2+} responses can discriminate between different downstream signalling pathways.

These different Ca^{2+} sensors also have distinct Ca^{2+} thresholds. By simply adding increasing amounts of calcium ionophore one can induce progressively larger Ca^{2+} responses. At low Ca^{2+} levels, such as those seen in tolerant B cells, NF-AT but not NF-κB or JNK is activated. At higher Ca^{2+} levels, such as those reached during the Ca^{2+} peak in naïve B cells, all three pathways are activated. Thus, the magnitude of the Ca^{2+} response, in addition to the duration of the response, can distinguish between downstream signalling pathways.

In summary, these findings demonstrate that maintenance of self-tolerance requires persistent antigen exposure. The signals induced by autoantigens in tolerant cells are qualitatively different from the signals induced by foreign antigens in naïve cells. Quantitative differences in the magnitude of signalling through second messengers such as Ca^{2+} contribute to this persistent but selective signalling. The two phases of the Ca^{2+} response serve different purposes (Fig. 4). The transient spike is stimulated by foreign antigen during an immune response and arises from internal and external Ca^{2+} sources. This brief high-magnitude signal results in sustained NF-κB and JNK but not NF-AT activation. The sustained plateau requires external Ca^{2+} and activates NF-AT. Genetic and pharmacological lesions which affect the magnitude or duration of the Ca^{2+} response will alter the activation of NF-κB, JNK and NF-AT and tamper with the balance between protective immunity and autoimmunity.

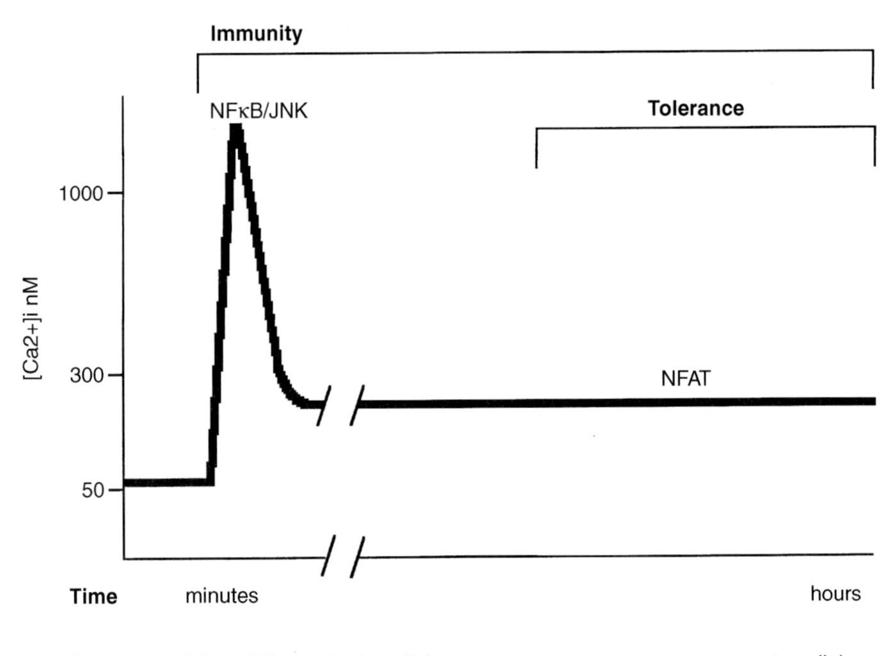

FIG. 4. Schematic illustration of the source, magnitude, duration and pathways of signals that are activated by each phase of the Ca^{2+} response. Foreign antigen stimulates a biphasic Ca^{2+} flux in naïve B lymphocytes while self antigen causes low level Ca^{2+} oscillations whose mean resembles the Ca^{2+} plateau. The transient Ca^{2+} peak is derived from intracellular and extracellular Ca^{2+} sources and activates NF-κB and JNK. The sustained plateau depends on extracellular Ca^{2+} and is required to keep NF-AT active.

References

Alferink J, Schittek B, Schönrich G, Hämmerling GJ, Arnold B 1995 Long life span of tolerant T cells and the role of antigen in maintenance of peripheral tolerance. Int Immunol 7:331–336

Cooke MP, Heath AW, Shokat KM et al 1994 Immunoglobulin signal transduction guides the specificity of B cell–T cell interactions and is blocked in tolerant self-reactive B cells. J Exp Med 179:425–438

Cyster JG, Goodnow C 1995 Antigen-induced exclusion from follicles and anergy are separate and complementary processes that influence peripheral B cell fate. Immunity 3:691–701

Cyster JG, Hartley SB, Goodnow CC 1994 Competition for follicular niches excludes self-reactive cells from the recirculating B-cell repertoire. Nature 371:389–395

Dolmetsch R, Lewis R, Goodnow C, Healy J 1997 Differential activation of transcription factors induced by Ca^{2+} response amplitude and duration. Nature 386:855–858

Fulcher DA, Basten A 1994 Reduced life span of anergic self-reactive B cells in a double-transgenic model. J Exp Med 179:125–134

Gajewski TF, Lancki DW, Stack R, Fitch FW 1994 Anergy of TH0 helper T lymphocytes induces down regulation of TH1 characteristics and a transition to a TH2-like phenotype. J Exp Med 179:481–491
Goodnow CC, Brink R, Adams E 1991 Breakdown of self-tolerance in anergic B lymphocytes. Nature 352:532–536
Hartley SB, Cooke MP, Fulcher DA et al 1993 Elimination of self-reactive B lymphocytes proceeds in two stages: arrested development and cell death. Cell 72:325–335
Healy JI, Dolmetsch RE, Timmerman LA et al 1997 Different nuclear signals are activated by the B cell receptor during positive versus negative signalling. Immunity 6:419–428
Höllsberg P, Weber WE, Dangond F et al 1995 Differential activation of proliferation and cytotoxicity by an altered peptide ligand. Proc Natl Acad Sci USA 92:4036–4040
Jenkins MK, Schwartz RH 1987 Antigen presentation by chemically modified splenocytes induces antigen-specific T cell unresponsiveness *in vitro* and *in vivo*. J Exp Med 165:302–319
Lamb JR, Skidmore BJ, Green N, Chiller JM, Feldmann M 1983 Induction of tolerance in influenza virus-immune T lymphocyte clones with synthetic peptides of influenza hemaglutinin. J Exp Med 157:1434–1447
Mueller DL, Chiodetti L, Bacon PA, Schwartz RH 1991 Clonal anergy blocks the response to IL-4, as well as the production of IL-2, in dual-producing T helper cell clones. J Immunol 147:4118–4125
Otten GR, Germain RN 1991 Split Anergy in a CD8+ T cell: receptor dependent cytolysis in the abscence of interleukin-2 production. Science 251:1228–1231
Rathmell JC, Cooke MP, Ho WY et al 1995 CD95(Fas/APO-1)-dependent elimination of self-reactive B cells upon interaction with CD4+ T cells. Nature 376:181–184

DISCUSSION

Mitchison: Is that biochemical phenotype one which enables you to look in the normal animal for anergized or tolerant cells?

Healy: Experiments addressing this have been delayed by the relocation of the lab to Australia. By assaying IgM density and the signalling phenotype I just described one could search for autoreactive B cells *in vivo*.

Jenkins: How will you know they are autoreactive?

Healy: The best way is to sort B cells which have low level calcium signalling and low IgM density, and then treat them with LPS and look at autoantibody production in these potentially autoreactive B cells.

Stockinger: Is there a good experiment on the T cell side? It was shown in the B cell system that anergic cells are short-lived—this explains a lot. On the T cell side, there is a lot of controversy and Bernd Arnold even proposes that there is a kind of memory cell for anergy. The experiments that have been done trying to take anergic cells out of the presence of self antigen and showing they recover, all have the problem that these were not truly monospecific T cells and there could have been other explanations. What is the feeling about the longevity of an anergic T cell?

Also, why would Marc Jenkins still call his cells tolerant, although he says six weeks later when he gives the antigen again they then respond? I would have thought that's the perfect exclusion of anergy.

Jenkins: They are tolerant during the period of time when the antigen is present. When the antigen is not present there's no need to be tolerant anymore.

Mitchison: But with regard to the first part of your question, is there anything like a corresponding biochemically well-defined anergic phenotype of T cells? Mike Lenardo spent several years of his life investigating this issue.

Lenardo: A group in Sweden recently published results on staphylococcal enterotoxin B (SEB)-treated mice (Sundstedt et al 1996) that were similar to our older data on T cell clones (Kang et al 1992) and showed that alterations in the AP-1 transcription factor caused a deficit in the production of IL-2.

Jenkins: I would say that experiment has not been done, because until recently we haven't had very good *in vivo* models.

Allen: Jim Healy, have you tried cross-linking using different forms of the antigen? You are giving monovalent antigen: do you change your phenotype with multimerized antigen?

Healy: Yes, Srini Akkaraju and I have data showing that dimeric antigen gives a strikingly different outcome compared with monomeric antigen. As of yet we have not tested beyond dimers.

References

Kang SM, Beverly B, Tran AC, Brorson K, Schwartz RH, Lenardo MJ 1992 Transactivation by AP-1 is a molecular target of T cell clonal anergy. Science 257:1134–1138

Sundstedt A, Sigvardsson M, Leanderson T, Hedlund G, Kalland T, Dohlsten M 1996 *In vivo* anergized $CD4^+$ T cells express perturbed AP-1 and NF-κB transcription factors. Proc Natl Acad Sci USA 93:979–984

Tolerance induction with CD4 monoclonal antibodies

Herman Waldmann, Frederike Bemelman and Stephen Cobbold

Sir William Dunn School of Pathology, South Parks Road, Oxford OX2 7QY, UK

Abstract. One of the major goals of therapeutic immunosuppression is to be able to use short-term therapy to achieve long-term tolerance. Short courses of CD4 antibodies are able to create peripheral tolerance in a mature immune system. The resulting tolerant state shows evidence of being dominant in that one can observe the features of linked suppression, transferable suppression and infectious tolerance in a variety of model systems. Only in the situation of administration of high doses of marrow could one find evidence of central and peripheral tolerance which had all the features of being deletional rather than regulatory. These findings suggest that attaining dominant tolerance and linked suppression may be the least invasive of all tolerance-inducing strategies for clinical application.

1998 Immunological tolerance. Wiley, Chichester (Novartis Foundation Symposium 215) p 146–158

Immunological diseases (autoimmunity, hypersensitivity and transplant rejection) account for a significant expenditure of health-budget resources in developed countries, ranking third behind cancer and cardiovascular disease. Management is largely palliative rather than curative, requiring the long-term administration of anti-inflammatory and immunosuppressive drugs that penalize the whole immune system for the unwanted activities of a very small minority of lymphocytes. As a result, patients carry a long-term risk of infection and a variety of other drug-related side-effects.

The ideal immunosuppressants would be low impact agents that could yield long-term benefit from short-term treatment yet retaining selectivity just for the culpable antigens. This requires that we should be able to instate or restore immunological tolerance. As T cells are crucial to all significant forms of immune reactivity to tissue antigens, our goal should be to achieve tolerance in that population. In view of the limitations of the human thymus to function in adult life, T cell tolerance should, ideally, be achieved without wholesale ablation of the peripheral T cell population.

In 1985 we observed that some tissue transplants given under the therapeutic umbrella of lytic CD4 and CD8 monoclonal antibodies (mAbs) survived long-term well beyond the period of immune deficiency imposed by lymphocyte depletion (Cobbold & Waldmann 1986). In 1986 our group and that of David Wofsy demonstrated that short-term therapy with CD4 antibodies could result in long-term tolerance to antigens co-administered at the same time (Benjamin & Waldmann 1986, Benjamin et al 1986, Gutstein et al 1986). Both groups subsequently demonstrated that non-lytic forms of CD4 antibody could do the same (Carteron et al 1988, 1989, Qin et al 1987, 1989, 1990, Benjamin et al 1988). Not only could CD4 antibodies promote tolerance in a naïve immune system, they could also do so in one which was previously primed, or still actively responding. All these properties conformed to those of an 'ideal immunosuppressant'. For this reason we have attempted to establish the mechanisms involved in these forms of peripheral tolerance in the hope that this knowledge would shape the design of future immunosuppressive agents, permit the development of diagnostic tools for monitoring tolerance and perhaps shed new light on mechanisms underlying self-tolerance.

In a previous Ciba Foundation meeting on a similar theme we presented some of our initial data on the subject where we claimed that peripheral tolerance promoted by CD4 antibodies could operate independently of $CD8^+$ T cells. The reader is encouraged to look at the discussion following that chapter (Waldmann et al 1987) and to note with some 'humour' the dismay and alarm of some of the 'sages' of immunology at any claim that T cell suppression could be mediated by any T cell other than the $CD8^+$ T cell. Since that time we have accumulated substantive evidence for the existence of $CD4^+$ T cells mediating dominant tolerance reflecting their potent capacity to interfere with T cell-mediated tissue attack. This chapter will summarize our knowledge on how these cells might function to ensure stable tolerance long-term. All the data are derived from studies using skin, vascularized heart and bone marrow grafts in mice.

Evidence for dominant tolerance mediated by $CD4^+$ T cells

In all transplant models that we have studied (skin, heart, marrow), animals tolerant of donor tissue (A) as a result of non-lytic CD4 mAb therapy can be shown to demonstrate three features all reflecting dominant tolerance exerted by $CD4^+$ T cells, namely resistance, linked suppression and transferable suppression (Table 1). In the case of heart grafts it is remarkable that adoptive transfer of tolerant cells will impose tolerance on a normal mouse such that this secondary recipient will accept an MHC-mismatched heart graft (A), not only from the same donor, but also from an $(A \times B)F1$ hybrid between that donor and a third party (B). Further transfer of tolerant T cells from these acceptors of F1 hearts to

TABLE 1 Evidence for dominant tolerance mediated by CD4+ T cells

Observation	*Reference*
1. Difficulty breaking tolerance by infusion of naïve T cells.	Wise et al 1992, Scully et al 1994
2. This resistance is mediated by T cells from tolerant host.	Wise et al 1992, Scully et al 1994
3. Tolerance to graft A results in acceptance and tolerance of graft (A × B)F1(linked suppression).	Davies et al 1996a, Chen et al 1996
4. $CD4^+$ T cells from tolerant animals suppress naïve T cells on co-transfer. $CD8^+$ T cells do not.	Qin et al 1993, Davies et al 1996b, Bemelman et al 1998
5. $CD4^+$ T cells from primed, then tolerized, animals suppress both naïve and primed T cells on co-transfer. $CD8^+$ T cells do not.	Marshall et al 1996
6. $CD4^+$ T cells transferred from mice tolerized to MHC-mismatched heart grafts prevent naïve mice from rejecting hearts from the same donors.	Chen et al 1996
7. T cells from the above animals accepting (A × B)F1 hearts are able to suppress a second cohort of naïve mice to accept B-type hearts.	Chen et al 1996
8. Mice tolerized to low doses of 'minor'-mismatched marrow grafts (A) accept skin from (A × B)F1 donors (linked suppression).	Bemelman et al 1998
9. $CD4^+$ T cells from these mice transfer linked suppression to naïve mice.	Bemelman et al 1998

a tertiary recipient enables these recipient animals to accept heart grafts from B. As no prophylactic immunosuppression is provided for the recipients of (A × B)F1 or B grafts, we must conclude that the regulatory reactions responsible for the spread of tolerance are all 'naturally' endowed within the immune system and simply revealed/amplified by CD4 mAb therapy in the primary recipient.

How is dominant tolerance maintained life-long?

Once dominant tolerance has been imposed on an immune system, the question arises as to whether the first cohort of regulatory $CD4^+$ T cells maintain tolerance throughout life, or whether the role gets taken on by new generations of regulators.

This question has been answered in three models of tolerance induced by CD4 antibodies (Table 2), and clear evidence for 'infectivity' of the tolerance process from donor to host T cells has been obtained. In the latter case, infectious tolerance was accompanied by demonstrable linked-suppression showing that these events are coupled or indeed part of the same process (Bemelman et al 1998).

TABLE 2 Infectious tolerance

Evidence for infectious tolerance	*Marker*	*Reference*
Naïve T cells become tolerant when transferred into a tolerant host	Human CD2	Qin et al 1993
Tolerant T cells impose heart graft tolerance on a secondary naïve host	Thy1.1 vs. Thy1.2	Chen et al 1996
Tolerance can be transferred through 10 sequential transfers to secondary naïve hosts	Dilution factor suggests 'infectivity'	Chen et al 1996
Tolerant T cells from marrow-tolerant donors (on boosting) impose tolerance on secondary naïve hosts	Thy1.1 vs. Thy1.2	Bemelman et al 1998

The demonstration that tolerance is lost once T cells are parked away from antigen (Scully et al 1994, Marshall et al 1996), suggests that antigen must be available to continuously sustain the regulators and the generation of any new ones by infectious tolerance. This and the finding that CD4 depletion in primed tolerant hosts results in tolerance breakdown (Marshall et al 1996) suggest that at least some T cells with the potential to reject grafts remain intact in tolerant animals, yet are continuously subdued by a process that requires CD4 regulators and a continued source of antigen.

Is CD4 antibody-mediated tolerance always dominant?

We have described one model where CD4 antibody treatment gives rise to a form of tolerance in which deletion rather than suppression is the dominant feature. In one situation (AKR marrow into CBA/Ca recipients) high doses of marrow led to deletion of Vβ6 cells in the thymus and periphery whereas low doses produced no deletion. High dose tolerance was accompanied by permanent acceptance of AKR skin grafts, but no linked suppression nor any transferable suppression. In contrast, low dose tolerance had all the features of dominant tolerance described earlier.

Many other groups have now published on a range of antibodies that can induce tolerance without the need to deplete T cells. Very few of these groups have thoroughly searched for evidence of dominant tolerance. The pervading concept seems to be that the different antibodies interfere with secondary signals to T cells, and that T cells become tolerant by default (signal 1). Our own interpretation would be that in most cases the antibodies have created a selective cease-fire where the immune system cannot aggress the graft but can recognize its antigens so as to promote regulatory cells (Waldmann & Cobbold 1996). Once immunosuppression is withdrawn it would be the final ratio of regulators to effectors that determines whether the outcome is rejection or tolerance. We presume that conventional small-drug immunosuppressants are not so selective in their effects, and that is why they have not been successful agents for tolerance induction.

How is dominant tolerance achieved?

The way in which $CD4^+$ T cells suppress is unresolved, but we know that interleukin (IL)-4 is involved in one model (Davies et al 1996b), although we cannot demonstrate a role for IL-4 in another (Marshall et al 1996). A comprehensive search for evidence that Th2 cells are responsible for dominant tolerance provided no strong evidence for that proposal (Cobbold et al 1995). Nor is it clear that dominant tolerance is an active process. Understanding the cellular and molecular basis of these phenomena remains then a major challenge for the future. The linked-suppression studies tell us that regulator cells and the T cells they suppress, must somehow come into the same microenvironment, if not onto the surface of the same antigen-presenting cell.

In the design of future clarifying experiments one should not assume that the solution will come using the same kinds of technological approaches that have resolved mechanisms of positive immune mechanisms responses (e.g. helper function, CTLs). The finding of dominant tolerance need not imply that an active mechanism is responsible. The Civil Service model (Waldmann et al 1992) was proposed to make just that point, and to offer an explanation of how anergic, impotent or incompetent T cells might act to regulate the activities of other T cells.

Therapeutic application

The findings summarized above suggest that dominant tolerance can be imposed on an immune system, even one which has been previously primed. Such tolerance is peripheral, and can be life-long as a consequence of being infectious to other T cells generated by the host. Remarkably, dominant tolerance to a limited set of antigens can spread to other antigens if these are associated together on the same tissue. This could have enormous implications in xenografting if potential recipients could be tolerized to a universal antigen that could be engineered into all pig-derived organs for transplantation. In autoimmune disease it might be possible to collect an array of well-defined tissue antigens each selective to a particular organ-specific disease, with the intention that they be used as a means to extend tolerance to all other antigens in the tissue.

References

Bemelman F, Honey K, Adams L, Cobbold SP, Waldmann H 1998 Bone marrow transplantation induces either clonal deletion or infectious disease tolerance depending on the dose. J Immunol 160:2645–2648

Benjamin RJ, Waldmann H 1986 Induction of tolerance by monoclonal antibody therapy. Nature 320:449–451

Benjamin RJ, Cobbold SP, Clark MR, Waldmann H 1986 Tolerance of rat monoclonal antibodies: implications for serotherapy. J Exp Med 163:1539–1552

Benjamin RJ, Qin SX, Wise MP, Cobbold SP, Waldmann H 1988 Mechanisms of monoclonal antibody-facilitated tolerance induction: a possible role for the CD4 (L3T4) and CD11a (LFA-1) molecules in self–non-self discrimination. Eur J Immunol 18:1079–1088

Carteron NL, Wofsy D, Seaman WE 1988 Induction of immune tolerance during administration of monoclonal antibody to L3T4 does not depend on depletion of L3T4$^+$ cells. J Immunol 140:713–716

Carteron NL, Schimenti CL, Wofsy D 1989 Treatment of murine lupus with Fab′2 fragments of monoclonal antibody to L3T4. Suppression of autoimmunity does not depend on T helper cell depletion. J Immunol 142:1470–1475

Chen ZK, Cobbold SP, Waldmann H, Metcalfe S 1996 Amplification of natural regulatory immune mechanisms for transplantation tolerance. Transplantation 62:1200–1206

Cobbold SP, Waldmann H 1986 Skin allograft rejection by L3/T4$^+$ and Lyt-2$^+$ T cell subsets. Transplantation 41:634–639

Cobbold SP, Adams E, Marshall SE, Davies JD, Waldmann H 1995 Mechanisms of peripheral tolerance and suppression induced by monoclonal antibodies to CD4 and CD8. Immunol Rev 148:1–29

Davies JD, Leong LYW, Mellor A, Cobbold SP, Waldmann H 1996a T cell suppression in transplantation tolerance through linked recognition. J Immunol 156:3602–3607

Davies JD, Martin G, Phillips J, Marshall SE, Cobbold SP, Waldmann H 1996b T cell regulation in adult transplantation tolerance. J Immunol 157:529–533

Gutstein NL, Seaman WE, Scott JH, Wofsy D 1986 Induction of immune tolerance by administration of monoclonal antibody to L3T4. J Immunol 137:1127–1132

Marshall SE, Cobbold SP, Davies JD, Martin GM, Phillips JM, Waldmann H 1996 Tolerance and suppression in a primed immune system. Transplantation 62:1614–1621

Qin SX, Cobbold SP, Tighe H, Benjamin R, Waldmann H 1987 CD4 monoclonal antibody pairs for immunosuppression and tolerance induction. Eur J Immunol 17:1159–1165

Qin SX, Cobbold SP, Benjamin R, Waldmann H 1989 Induction of classical transplantation tolerance in the adult. J Exp Med 169:779–794

Qin SX, Wise M, Cobbold SP et al 1990 Induction of tolerance in peripheral T cells with monoclonal antibodies. Eur J Immunol 20:2737–2745

Qin SX, Cobbold SP, Pope H, Elliott J, Kioussis D, Waldmann H 1993 Infectious transplantation tolerance. Science 259:974–977

Scully R, Qin SX, Cobbold SP, Waldmann H 1994 Mechanisms in CD4 antibody-mediated transplantation tolerance: kinetics of induction, antigen dependency and role of regulatory T cells. Eur J Immunol 24:2383–2392

Waldmann H, Cobbold S 1996 How may immunosuppression lead to tolerance? The war analogy. In: Banchereau J, Dodet B, Schwartz R, Trannoy E (eds) Immune tolerance. Elsevier, Paris, p 221–227

Waldmann H, Cobbold SP, Qin S et al 1987 Monoclonal antibodies for the depletion of specific subpopulations of lymphocytes. In: Autoimmunity and autoimmune disease. Wiley, Chichester (Ciba Found Symp 129) p 194–208

Waldmann H, Qin S, Cobbold S 1992 Monoclonal antibodies as agents to reinduce tolerance in immunity. J Autoimmun 5:93–102

Wise M, Benjamin R, Qin S, Cobbold S, Waldmann H 1992 Tolerance induction in the peripheral immune system. In: Vogel H, Alt F (eds) Molecular mechanisms of immunological self-recognition. Academic Press Inc., Troy, MO, p 149–155

DISCUSSION

Stockinger: When you talk about this 10-fold sequential transfer, before each transfer do you challenge the donor with a skin graft and then take the cells for the next transfer?

Waldmann: No, the animals have had heart grafts. We then wait 100 days while the heart graft is still there, and then transfer those cells from that animal.

Stockinger: But in both heart and skin grafts, after a certain period there is no class II-expressing cell left. Wouldn't you call those antigen-free in terms of your CD4 regulatory population?

Waldmann: We also get tolerance to any antigens which are processed indirectly by the host. If you actually have a recipient (A × B)F1 and you now put on a graft that has a compatible (say A) MHC but differs in minors, you now become tolerant to those minors restricted by the other MHC haplotype (B). What this means is that whatever graft you have in there, there are a lot of antigens: there are those that presented directly and those that are processed, including MHC. I wouldn't know which were the antigens that were determining the tolerant state.

Arnold: In the latter part of your paper you said that depending on how many bone marrow cells you give, you have different kinds of tolerance. Could this therefore be a multistep process, with the cells bumping into bone marrow cells more often if you give high doses of bone marrow and therefore you get deletion? Or do you think it is a different place, because a lot of bone marrow cells go somewhere else and deletion occurs elsewhere?

Waldmann: I think both suggestions are reasonable. It's an important question to ask.

Arnold: Have you done this in thymectomized animals? Is the thymus needed for deletion?

Waldmann: We have not done that in this system because we wanted to look in the thymus, but we know from previous work that the thymus doesn't need to be present to get tolerance to the graft.

Kurts: With regard to therapy, if you plan to use these antibodies in clinical situations such as transplantation, you would tolerize the patients not only to the graft, but also to whatever pathogens are present at the time. For example, if the patient has a virus infection at this time then they will remain tolerant to this virus. Can you make your therapy a little more specific?

Waldmann: There are a lot of questions here. One concerns whether it is just as easy to tolerize to all antigens: this doesn't seem to be the case. I am amazed that it is so easy to tolerize transplantation antigens. From the very beginning Gary Fathman couldn't tolerize to KLH and we couldn't tolerize to ovalbumin: we've never understood why not.

There are many assumptions here. One concerns the sorts of issues we've discussed today about adjuvants and whether or not they are relevant to this situation. After all, we are not putting in antigens with adjuvant. Viruses may carry their own so-called 'danger signal' that may override some of our interventions. If you are starting off in a totally theoretical framework, you would say that tolerizing to pathogens is something we have to be careful of, but I don't think we know that we can tolerize to any live virus or bacteria. Having said that, I'm not interested in using CD4 antibodies to tolerize patients until we know better how they work.

Kurts: You could address these questions by infecting mice when you graft them and tolerize them to the graft using your therapeutic protocol. It would be interesting to find out whether they can clear the pathogen and develop protective immunity.

Waldmann: A lot of work was done in the early days looking at model systems of infection with ablative antibodies. The trouble with the viral system is that if you don't cure the infection the virus just proliferates making the readouts inadequate, so limiting the assessment of tolerance. It's a problem of immunosuppression: when you give cyclosporin you immunosuppress. This is the problem when you try to do the experiment you are proposing.

Abbas: If you set up an allo mixed lymphocyte reaction (MLR), especially with the MHC mismatched grafts that you are now able to do, and tolerize, what happens to T cells in the animals that are carrying the grafts?

Waldmann: We did a lot of work with tolerance to MHC mismatch skin grafts. Basically, we could not find *in vitro* evidence of tolerance. We got MLRs and CTL responses, and IFN-γ release. In the end we came to the conclusion that we could not find an *in vitro* correlate. But I am absolutely happy that we have got a great *in vivo* system.

Shevach: You are trying to avoid the issue of whether Th2 cells are responsible for aggression. I'm more impressed by your observation that anti-IL-4 reverses the suppression. A simple assay for the activity of Th2 cells would be antibody production. Is there antibody production in the tolerant animals, particularly with the major MHC difference, and what is the subclass of antibody?

Waldmann: When we tolerize to MHC mismatched grafts there's no antibody at all.

Shevach: Why haven't you pursued the blocking experiments? This seems to be the best possible situation to generate a Th2 cell, for the very arguments you actually gave: the mode of antigen presentation in these studies is atypical as the initial response in the graft is prevented. Antigens shed by the graft and represented may be highly effective inducers of a Th2 response.

Waldmann: A great discussion point. What is the evidence you would require to say a T cell is Th2?

Shevach: Make one partner an IL-4 knockout.

Waldmann: You can tolerize IL-4 knockouts.

If someone tells me what clearly defines a Th2, I can tell them whether I think we've got Th2 or not. Ethan Shevach, you put the question — you must have some idea.

Shevach: Can you get infectious tolerance in the IL-4 knockout? That's the key. We're not talking about regular tolerance, which is simple, but infectious tolerance.

Waldmann: Steve Cobbold has written a review of all the things we tried to do to examine cytokine profiles using limiting dilution and so on (Cobbold et al 1996). We can't see a big effect. It would be fickle to say it's not Th2, it would be fickle to say it is Th2: I don't know what criteria you guys want.

Shevach: We accept that nothing *in vitro* counts, so it's really *in vivo* treatment during those 100 days with anti-IL-4 that would allow you to conclude that Th2 cells are involved in suppression.

Mitchison: Isn't it the situation that you're not willing to conclude that you have a Th2 shift, even though you have a suggestive blocking experiment? But the Hancock group in Boston, also using treatment with anti-CD4 antibodies, do find a Th2 shift in T cells infiltrating long-surviving rat allografts (Mottram et al 1995).

Waldmann: It's one thing to demonstrate at a certain point in time that certain cytokines of a Th2 type are present in the context of an animal being tolerant. It is much harder to prove that it's the Th2 phenotype that is responsible for that tolerance.

Mitchison: You need the antibody blocking experiment.

Waldmann: We can't block induction of tolerance with anti-IL-4. We can block transferable suppression with anti-IL-4, but when we transfer tolerant primed cells that suppress we cannot block with anti-IL-4. When we use TCR transgenics and can get intracellular staining, maybe we can give you a fair conclusion. I would actually say we are doing these experiments and we want to interpret our results in an open way: why can't we keep everything open? Why the hassle?

Mitchison: But that is a dangerous position to be in when other people maintain that they have cracked the problem. That is the position that Hancock and Carpenter have taken in the work cited above.

Waldmann: Terry Strom, part of the same group, is now saying that the Th1/Th2 paradigm may not completely explain the data.

Jenkins: Have you looked in the grafts that are bearing the antigen to which suppression is directed? Is there any infiltrate and are there any eosinophils?

Waldmann: No.

Jenkins: So you don't really know then whether it has really not infiltrated or whether there is a different type of infiltrate.

Waldmann: We know early on there is an infiltrate in quite a few of the grafts, but we don't know whether it persists.

Mitchison: Is the infiltrate IL-4 positive?

Waldmann: We haven't looked.

Jenkins: It might be interesting to look for a non-pathogenic infiltrate. This might tell you something about what's going on. If you prime a lot of Th2 cells, you will have eosinophils in the infiltrate.

Waldmann: Many other people who have looked with our antibodies at heart grafts and other things have not seen pathogenic infiltrates.

Lechler: I would like to come back to the question of what maintains the CD4 regulatory population. Possibly, your answer is problematic. There seem to be three possibilities. The first is that it is the keratinocytes in the graft, and if there is some grumbling inflammation and class II expression maintained, that would be fine. Another possibility, which Tom Starzl and his colleagues would love, is that some of the Langerhans' cells persist after emigrating from the graft. These would both present donor class II and maintain tolerance in your CD4 population. If your suggestion is correct—that it is CD4 cells with indirect specificity for the graft—then I have a problem anatomically in envisaging how they would regulate CD8 cells that presumably have direct specificity for class I of the graft. So if you want your three-cell model, it won't work.

Waldmann: I agree. I can imagine a situation where we can get a three-cell model to deal with linked suppression across an MHC difference. The reprocessing issue is a problem: it actually implies that somehow there must be a greater bystander-type reaction than that.

Lechler: Perhaps there is some persistent class II expression on the graft: I wouldn't rule that out at all. Certainly, if you have got a little bit of infiltration going on, a little bit of IFN-γ, then it may well be there's persistence of donor antigen.

Mitchison: Could we bring in another body of relevant information, which is summarized in a review by Kim Bottomly (Constant & Bottomly 1997). In experiments with anti-OVA transgenic cells, and in other experiments with type IV human collagen in mice, different forms of TCR signal attenuation—for example, reducing antigen concentration using an inappropriate MHC molecule or blocking co-stimulation—all drive Th2 differentiation. Don't your findings fit admirably into that context?

Waldmann: Admirably? I don't know.

Abbas: Even before you deal with that, not everyone finds that low-dose peptide or the mutant peptide gives Th2 responses. In fact, most of the data come to the opposite conclusion. The CD28 knockout mouse is a Th1 mouse (Rulifson et al 1997). Low dose antigen tends to induce Th1 responses; one can cure a BALB/c mouse by reducing the dose of *Leishmania* down two logs because it now stimulates Th1 responses (Bretscher et al 1992).

Shevach: I would actually claim that in this model you have high-dose antigen presentation for the very reason Robert Lechler just described. There is simultaneous induction of IFN-γ which results in prolonged expression of high levels of MHC antigens. This may result in induction of Th2 cells.

Mitchison: But you wouldn't deny that blocking CD4 is a signal-attenuating strategy, would you?

Shevach: It's surely attenuating the initial signal for T cell activation and prevents rejection of the graft. What happens after that is much more complicated.

Hämmerling: You said that there is a high dose antigen. However, the graft is very small. There is probably a high local concentration but the total amount is extremely low.

Shevach: The T cells sees the high local level of antigen.

Lechler: Given that we have a chairman who is so enthusiastic about the Th1/Th2 hypothesis, and we have a speaker who coined the term 'the civil servant model of suppression', I would be interested to know whether you think it is a possible player, since you first proposed the idea and you know our *in vitro* data which fulfil every prediction that you would care to make.

Waldmann: Before we had done the analysis fully, we published a paper where we speculated that it was Th1 shifting to Th2 (Waldmann & Cobbold 1993). We looked very hard on the local lymph node after a fresh graft. We did as much as we could to see that switch and we couldn't see it. I don't want to be forced into saying it must be Th2 or it is not Th2; we just don't have the information. We have kept a completely open mind.

There must be many cells in the immune system that remain after these various antigen exposures that see antigen but are not able to bring about destructive inflammation. The question is, are those cells garbage or are they junk? Garbage you throw away, junk you want to keep. In my mind they are junk. Is it possible that in these procedures, when we generate cells which are incapacitated and cannot go right through to the final damaging stage, they in some way can block at sites of antigen presentation. I called that the 'civil service' model. The idea is that the whole immune system works on cooperation between T cells. No one has ever described a naïve cell starting from scratch and producing an immune response. The way the field has gone is that if you have a T cell seeing antigen in isolation, it might become tolerant too. The civil service model was to say that if there were a lot of incapacitated cells and they were all coming to a site of antigen presentation, a cell that was uncommitted and looking for collaboration would not find it. Therefore it would become anergic and the whole thing, in a passive way, could cascade.

Shevach: But how do they teach? I can see how they can compete.

Waldmann: They don't teach. Here's a cell that's uncommitted: it comes into an environment where it cannot get collaboration. It becomes like them, it accumulates. The next one comes out, it accumulates. We don't have to have an active mechanism.

Lenardo: I don't understand the molecular justification for this paradigm. As far as I know, as long as a T cell sees an antigen-presenting cell (APC) that gives it all the right signals, it is capable of being activated on its own.

Waldmann: A T cell clone or a naïve T cell?

Lenardo: A naïve T cell.

Waldmann: Who has ever done that?

Lenardo: If you have a clone, or a transgenic T cell that is molecularly well-defined, you can do this.

Hafler: You need T–T collaboration to get clonal expansion.

Abbas: But you must have done limiting dilution-type analyses with TCR transgenics.

Waldmann: Mark Jenkins, you did that with Ron Schwartz in the early days and you never got down to a single clone.

Jenkins: That was Dan Mueller's work. He did show that single cloned T cells failed to respond to anti-TCR stimulation. However, no source of co-stimulation was present, which would not be the case when a single antigen-specific T cell interacts with an APC in a lymph node.

Shevach: Why would it work? That is what Abul is asking.

Lechler: Paracrine stimulation. It's an assumption. We know that T cells cluster, and that's what dendritic cells are probably particularly good at doing, so it is an assumption that T cells need paracrine stimulation in order to clonally expand enough to remain 'on'. But it's a very difficult assumption to test.

References

Bretscher PA, Wei G, Menon JN, Bielefeldt-Ohmann H 1992 Establishment of stable, cell-mediated immunity that makes 'susceptible' mice resistant to *Leishmania major*. Science 257:539–542

Cobbold SP, Adams E, Marshall SE, Davies JD, Waldmann H 1996 Mechanisms of peripheral tolerance and suppression induced by monoclonal antibodies to CD4 and CD8. Immunol Rev 149:5–33

Constant SL, Bottomly K 1997 Induction of Th1 and Th2 $CD4^+$ T cell responses: the alternative approaches. Annu Rev Immunol 15:297–322

Mottram PL, Han WR, Purcell LJ, McKenzie IF, Hancock WW 1995 Increased expression of IL-4 and IL-10 and decreased expression of IL-2 and interferon-γ in long-surviving mouse heart allografts after brief CD4-monoclonal antibody therapy. Transplantation 59:559–565

Rulifson IC, Soeling AI, Fields PE, Fitch FW, Bluestone JA 1997 CD28 costimulation promotes the production of Th2 cytokines. J Immunol 158:658–665

Waldmann H, Cobbold SP 1993 The use of monoclonal antibodies to achieve immunological tolerance. Trends Pharmacol Sci 14:143–148

A two-step model for the induction of organ-specific autoimmunity

Andreas Limmer, Torsten Sacher, Judith Alferink, Thomas Nichterlein*, Bernd Arnold and Günter J. Hämmerling

*Division of Molecular Immunology, Tumor Immunology Program, German Cancer Research Center, Im Neuenheimer Feld 280, 69120 Heidelberg, and *Institute of Medical Microbiology and Hygiene, Faculty of Clinical Medicine Mannheim, University of Heidelberg, Germany*

Abstract. Peripheral tolerance is considered to be a safeguard against autoimmunity but the mere existence of anergic T cells renders them potentially dangerous. Using transgenic mice that were tolerant to a foreign MHC class I antigen (K^b) exclusively expressed in the liver, we investigated whether reversal of tolerance *in vivo* would directly result in autoimmunity. Breaking of tolerance was achieved by application of tumour cells expressing both K^b and interleukin 2. Despite the fact that the respective mice were now able to reject K^b-positive grafts, the reversed T cells did not infiltrate and attack the K^b-positive liver. However, when the liver was 'conditioned' through an inflammatory reaction either by irradiation or by infection with *Listeria*, massive T cell infiltration and liver damage were observed in the reversed mice. The results show that at least two steps are required for autoimmunity: (1) activation of antigen-specific T cells, and (2) conditioning of the target organ. It will be important to determine the factors leading to conditioning but it is likely that adhesion molecules are involved. These experiments are not only of relevance for treatment of autoimmune diseases but also for tumour therapy.

1998 Immunological tolerance. Wiley, Chichester (Novartis Foundation Symposium 215) p 159–171

Physical elimination of self-reactive thymocytes by apoptosis is a major mechanism for the establishment of self-tolerance (reviewed in Kisielow & von Boehmer 1995). In addition to this central tolerance there is also a need for peripheral tolerance mechanisms because (1) deletion in the thymus is a stochastic event which is unlikely to eliminate all autoreactive T cells, and (2) many antigens probably exist in the periphery that are not present in the thymus (e.g. tissue-specific antigens) for which induction of tolerance is necessary for the prevention of dangerous autoaggression.

Transgenic mice with organ-specific expression of antigens outside the thymus have indeed demonstrated the existence of peripheral tolerance (reviewed in Hämmerling et al 1993, Arnold et al 1993). The respective mechanisms include:

(1) peripheral deletion of self-reactive cells which can be achieved by injection of superantigen or high doses of peptide; (2) induction of unresponsiveness or anergy, either directly or by immune deviation; or (3) the T cells are unresponsive because they ignore the antigen in the periphery. In agreement with this, numerous studies in rodents and humans have reported the existence of T cells which can easily respond *in vitro* against a large variety of self antigens (e.g. myelin basic protein, MBP). Such cells can be found in almost every normal individual (see Wekerle et al 1996) but the occurrence of autoimmunity is, fortunately, a rare event. The existence of these self-reactive cells marks them as prime candidates for the T cells that eventually cause autoaggression and autoimmunity, but the events leading to autoaggression are not clear. In the present report we describe that the induction of autoimmunity against the liver does not easily come about because at least two distinct steps are required: (1) breaking of tolerance in the periphery by activation of self-reactive (liver-specific) T cells; and (2) conditioning of the liver by an inflammatory response which will render it susceptible to attack by the activated self-specific T cells.

Peripheral tolerance in mice expressing K^b on hepatocytes

In the work described here we have investigated whether conditions could be established leading to reversal of peripheral tolerance *in vivo* and whether this would result in autoimmunity against the liver. To this end we used double-transgenic mice expressing a foreign MHC class I K^b molecule exclusively outside the thymus on hepatocytes due to the use of the liver-specific albumin or C-reactive protein (CRP) promoters. In addition, the mice were transgenic for a T cell receptor (Des.TCR) with specificity for the K^b molecule and which recognizes K^b only when it carries a particular peptide (A.-M. Schmitt-Verhulst, personal communication). The rationale for using a class I molecule as an autoantigen is that MHC class I molecules present peptides from endogenously synthesized antigens and that the MHC class I/peptide complex is not supposed to travel to other cells in a form that can still be recognized by T cells. The main characteristics of the peripheral tolerance in these mice have been described previously (Schönrich et al 1991, 1992, Ferber et al 1994) and can be summarized as follows: Des.TCR-positive CD8 cells are present in the periphery of double-transgenic mice but fail to reject K^b-positive grafts such as skin or tumour grafts, whereas Des.TCR single-transgenic mice efficiently reject these grafts. Both, the Des.TCR × Alb.K^b and Des.TCR × CRP.K^b mice display down-regulation of their Des.TCR on a fraction of CD8 cells, but there are also substantial numbers of CD8 cells with normal Des.TCR expression. However, after enhancement of K^b expression on hepatocytes achieved by induction of the CRP promoter with lipopolysaccharide (LPS), a more pronounced down-regulation of the Des.TCR was observed, indicating that the tolerant cells were not totally inert but still responsive to an antigenic stimulus, although in a negative manner (Ferber et al 1994). These findings suggest that the non-deletional peripheral tolerance observed reflects a dynamic state that is influenced by the presence and amount, and possibly also the nature, of the antigen (Alferink et al 1995). Interestingly, in the absence of LPS stimulation the level of K^b expression on hepatocytes in CRP.K^b mice was extremely low and only detectable by PCR, suggesting that almost undetectable amounts of antigen can be sufficient to induce tolerance in the periphery (Ferber et al 1994). This finding made it necessary to rule out that the observed tolerance was induced by tiny amounts of K^b in the thymus which we might have failed to detect. However, chimeric CRP.K^b transgenic mice whose thymus was replaced by a thymus from an H-2^k mouse, which does not express K^b, were also tolerant demonstrating that we are indeed confronted with peripheral tolerance induction events (Ferber et al 1994).

Breaking of tolerance

It has been reported that the tolerant status of T cell clones which had been rendered anergic by stimulation with ConA or fixed antigen-presenting cells (APCs) could be reversed *in vitro* with interleukin (IL)-2 (Beverly et al 1992). Therefore, we investigated whether or not IL-2 would be helpful in breaking tolerance *in vivo*. To this end, we used P815 (H-2^d) tumour cells and constructed double transfectants expressing both the K^b molecule and IL-2. These P815.K^b.IL-2 tumour cells were injected into Des.TCR $\times$ Alb.K^b or Des.TCR $\times$ CRP.K^b mice. In contrast to the P815.K^b tumour cells which readily grow in the tolerant mice, the P815.K^b.IL-2 failed to grow. When these mice were challenged several weeks later with P815.K^b cells, rejection was observed. These results indicate that by treatment with cells expressing both K^b and IL-2, tolerance could be reversed leading to the activation of K^b-reactive T cells that were able to reject K^b-positive grafts (our unpublished observations). The reaction was specific for K^b as demonstrated by the observation that the ability to reject P815.K^b could be abrogated by *in vivo* treatment with Des.TCR clonotypic antibody. Comparable observations were made in Des.TCR $\times$ Alb.K^b mice bred to the Rag-2 knockout background. Since such mice harbour only clonotypic CD8 cells, a contribution from other T cells such as CD4 can be ruled out. Breaking of tolerance resulted also in the appearance of markers on Des.TCR^+ CD8 cells characteristic of the activated phenotype, namely an increase of α_4 integrin, LFA-1, ICAM-1 and CD44, a transient up-regulation of CD25 (IL-2R), and down-regulation of L-selectin (our unpublished work).

Further studies showed that tolerance could not be reversed by injection of P815 cells expressing only IL-2 (P815.IL-2). Together, these observations suggest that tolerance can most efficiently be broken when antigen and IL-2 are applied in the close vicinity of each other. The most interesting result was, however, that these reversed T cells failed to attack the K^b-positive hepatocytes although they were clearly able to reject K^b-positive tumour grafts. This was documented by the failure of the reversed T cells visibly to infiltrate the liver and cause release of the liver-specific enzyme alanine amino transferase (ALT).

Reversed T cells can attack K^b-positive hepatocytes in irradiated recipients

The failure of the reversed T cells to cause liver damage could have been due to an intrinsic inability on their part to destroy hepatocytes. Therefore, the reversed T cells were transferred into irradiated Alb.K^b recipients that express K^b only on hepatocytes but nowhere else in the body. The results are summarized in Table 1. It can be seen that transfer of T cells into irradiated recipients always resulted in strong infiltration into the liver, regardless of the activation status of the T cells. However, damage of hepatocytes as documented by ALT release was only found when the T cells had been reversed, and not when resting T cells from either normal or tolerant mice were used, nor when the liver was K^b-negative. Transfer of reversed T cells and of T cells activated in Des.TCR single-transgenic mice by P815.K^b.IL-2 led to comparable amounts of ALT release, indicating that reversal of T cells was quite efficient.

These results show that the reversed T cells are, in principle, able to react against the K^b-positive hepatocytes, thus raising the question as to why this occurs only in

Table 1 Liver damage after transfer of activated T cells

Transferred cells	*K^b expression on liver in irradiated recipients*	*Infiltration of Des.TCR T cells*	*Liver damage (ALT release)*
Des.TCR, resting	K^b-positive	+++	—
Des.TCR, activated	K^b-positive	+++	+++
Des.TCR, activated	K^b-negative	+++	—
Tolerant Des.TCR, resting	K^b-positive	+++	—
Tolerant Des.TCR, activated	K^b-positive	+++	+++
Tolerant Des.TCR, activated	K^b-negative	+++	—

T cells from Des.TCR single transgenic or from tolerance Des.TCR × Alb.K^b mice activated two weeks before *in vivo* by P815.K^b.IL-2, or resting T cells, were transferred into irradiated recipients expressing K^b in the liver (Alb.K^b mice) or not (H-2^{dxk}) mice. Infiltration of Des.TCR T cells was evaluated by immunohistology, liver damage by determination of ALT release that ranged between 25 u/l background (−) and maximal release, around 200 u/l (+++).

irradiated recipients and not in the animal where the T cells had been activated. Irradiation is known to cause an inflammatory reaction characterized by the up-regulation of a variety of cytokines, chemokines, adhesion molecules, etc. Therefore, we attempted to see whether an inflammatory response in the mice, where the T cells had been reversed, would allow the T cells to attack the liver.

Infection with *Listeria* leads to autoimmunity after breaking of tolerance

Listeria monocytogenes is a Gram-positive facultative intracellular bacterium infecting mainly macrophages and hepatocytes. Activation of T cells by *Listeria* is usually maximal on day 5–6 as also documented by maximal ALT release during this period. Tolerant Des.TCR × Alb.K^b mice were infected i.v. with 10^4 colony forming units of *Listeria*, a dose which is high enough to cause T cell infiltration into the liver but too low to strongly activate T cells and result in significant ALT release. In agreement with this, injection of *Listeria* into the tolerant mice caused release of only small amounts of ALT above background level. However, when tolerance was first broken in Des.TCR × Alb.K^b mice by injection of P815.K^b.IL-2 cells and then the liver was infected with the *Listeria* one or two weeks later, strong release of ALT was observed with a maximum on day 2, followed by a gradual decline. Activation of T cells by P815.K^b.IL-2 in Des.TCR single-transgenic mice (that lack K^b in the liver) followed by *Listeria* infection did not result in significant ALT release. Likewise, TCR × CRP.K^b mice pre-treated with P815.IL-2 followed by infection did not exhibit significant ALT release. These observations are summarized in Table 2. The results show that histologically detectable infiltration into the liver and ALT release is only observed after reversal of T cell tolerance and after infection with *Listeria*. To rule out the possibility that this could be due to the high frequency of K^b-specific T cells in the transgenic mice, we repeated the experiments in Alb.K^b single-transgenic mice. Comparable results were obtained: pre-treatment with P815.K^b.IL-2 followed by *Listeria* infection led to massive ALT release on day 2, whereas no significant liver damage was observed when the mice were only pre-treated with P815.IL-2 (not shown).

Discussion and conclusions

The data presented here demonstrate that two steps are required for liver-specific autoimmunity: the activation of T cells and the conditioning of the target organ by an inflammatory response. In our experiments we activated the T cells by breaking tolerance with simultaneous application of antigen and IL-2. However, it is conceivable that in other systems, e.g. in the case of ignorant T cells, other

Table 2 Organ-specific autoimmunity after activation of K^b-specific T cells and conditioning (inflammation) of the liver

K^b-specific cells	*K^b-positive liver*	*Infiltration of Des.TCR T cells*	*Liver damage (ALT release)*
Non-activated	Non-conditioned	—	—
Activated	Non-conditioned	—	—
Non-activated	Conditioned	+++	—
Activated	Conditioned	+++	+++
Activated	K^b-negative liver, conditioned	+++	—

T cells in tolerant Des.TCR × Alb.K^b or Des.TCR × CRP.K^b mice were reversed (activated) *in vitro* by i.v. infection of P815.K^b.IL-2. One to two weeks later the liver was conditioned by infection with 10^4 *Listeria*. Note that strong infiltration and liver damage (ALT release) was only seen when tolerance was broken *and* the liver was conditioned by infection.

mechanisms might lead to activation. Importantly, the activated T cells appeared to be harmless as they failed to attack the liver.

It is not yet clear which factors render the liver susceptible to attack by T cells: this is why we prefer to use the term 'conditioning'. So far, most attempts to interfere therapeutically with autoimmunity have concentrated on the T cell side, e.g. antagonistic peptides that might interfere with antigen presentation and recognition, regulation of T cells or inactivation by T cell-specific antibodies. The present data strongly indicate that the target organ itself, or the 'conditioning' of it, may be an alternative target for therapeutic intervention (see also Steinmann 1996). Therefore, it will be of prime importance to delineate the precise factors leading to conditioning of the target organ. These could include cytokines and chemokines but the observation that massive infiltration becomes visible only after infection or irradiation suggests that adhesion molecules might also be involved. Indeed, after irradiation or infection with *Listeria* an up-regulation of adhesion molecules such as P- and L-selectin, ICAM-1 and notably VCAM-1 was observed in the liver by immunohistology. Up-regulation of VCAM-1 on hepatocytes after stimulation with LPS (van Oosten et al 1995) and of a variety of cytokines, adhesion molecules and chemokines after IL-2 treatment (Anderson et al 1996) has also been reported.

Adhesion molecules are known to be crucial for adherence of leukocytes and their transmigration through endothelial walls, a mechanism by which leukocytes usually find their way into those tissues that are normally shielded off. However, the liver parenchyma is supposed to be accessible for leukocytes. It is therefore puzzling that infiltration by activated T cells is only observed after inflammation. One explanation might be that normal access is possible but limited. Elevated expression of VCAM-1 (and other molecules) on both endothelial cells and hepatocytes may not only lead to enhanced transmigration but also to trapping of the activated T cells by the hepatocytes, allowing the T cells to exert their damaging function. In this context, it is encouraging to note that in MBP-induced experimental autoimmune encephalomyelitis (EAE), treatment with antibodies against VLA-4 led to an amelioration of the disease (Yednock et al 1992).

Another important message inherent in the data presented here is that once T cells are activated, autoimmunity can be induced by an organ-specific but antigen non-specific infection. Thus, there does not appear to be a need for mimicry. Formally, it cannot be excluded that *Listeria* might contain a K^b-binding epitope that is cross-reactive with the one recognized by the Des.TCR T cells, but the fact that comparable observations were made in irradiated mice render this possibility highly unlikely and less relevant (compare Tables 1 and 2). Obviously, this does not argue against the possible relevance of mimicry in other systems. In fact, a cross-reactive pathogen might not only activate self-reactive T cells but it might simultaneously perform the second step postulated in our model, namely the conditioning of the target organ through an inflammatory response.

It is important to note also that IL-2 alone was not sufficient to break tolerance and to cause autoimmunity, at least not in the amount secreted by the P815.IL-2 cells. Several years ago when reversal of clonal anergy *in vitro* by IL-2 alone was reported it was felt that such a mechanism would be too dangerous for the *in vivo* situation because the almost ubiquitous IL-2 might inadvertently reverse autoreactive cells *in vivo* thereby leading to dangerous autoaggression (see Schwartz et al 1996). The present data suggest that the situation is indeed not overly dangerous because reversal was only observed by application of antigen and IL-2 together. While the activated T cells are not directly autoaggressive, they still constitute a potential source of danger once the target organ is conditioned.

The two-step model described here may have certain similarities with the checkpoints in the progression of autoimmune diabetes described in NOD mice (reviewed in André et al 1996). There is a first checkpoint at three weeks of age that controls infiltration of T cells into the pancreas whereas checkpoint 2 at about 10–25 weeks of age is suspected to switch on the pathogenic potential of lymphocytes. There is a resemblance to the two-step model presented here but the sequence of events is reversed: here, the T cells were first activated (and thereby endowed with pathogenic potential) and thereafter infiltration was achieved by an inflammatory reaction. In this context, it should be emphasized that in experimental models of autoimmunity quite often the antigen itself is not sufficient for induction but requires an adjuvant, for example pertussis toxin in the case of EAE.

Finally, the two-step model might also be of relevance to tumour therapy. So far, most work on immunotherapy of tumours concentrates on the T cell side, namely activation of T cells, vaccine development, isolation and expansion of tumour infiltrating T cells, etc. However, it appears worthwhile to devote some effort to the tumour target itself and condition it in such a way that it becomes susceptible to attack by tumour-specific T cells. This might be of particular importance for tumours that do not contain an inflammatory infiltrate. This aspect re-emphasizes the importance of a precise definition of the molecular factors leading to the status of a target organ that we term 'conditioning'.

Acknowledgements

We thank Mrs Copson for the preparation of the manuscript and A. Klevenz, M. Kretschmar, G. Schönrich for their experimental support. This work was in part supported by the Biotechnology Programme BIO4-CT96–0077, G.J.H./B.A. and DFG Ar 152/3–1, B.A.

References

Alferink J, Schittek B, Schönrich G, Hämmerling GJ, Arnold B 1995 Long life span of tolerant T cells and the role of antigen in maintenance of peripheral tolerance. Int Immunol 7:331–336

Anderson JA, Lentsch AB, Hadjiminas DJ et al 1996 The role of cytokines, adhesion molecules, and chemokines in interleukin-2-induced lymphocytic infiltration in C57BL/6 mice. J Clin Invest 97:1952–1959

André I, Gonzalez A, Wang B, Katz J, Benoit C, Mathis D 1996 Checkpoints in the progression of autoimmune disease: lessons from diabetes models. Proc Natl Acad Sci USA 93:2260–2263

Arnold B, Schönrich G, Hämmerling GJ 1993 Multiple levels of peripheral tolerance. Immunol Today 14:12–14

Beverly B, Kang S-M, Lenardo MJ, Schwartz RH 1992 Reversal of *invitro* T cell clonal anergy by IL-2 stimulation. Int Immunol 4:661–671

Ferber I, Schönrich G, Schenkel J, Mellor A, Hämmerling GJ, Arnold B 1994 Levels of peripheral T cell tolerance induced by different doses of tolerogen. Science 263: 674–676

Hämmerling GJ, Schönrich G, Ferber I, Arnold B 1993 Peripheral tolerance as a multi-step mechanism. Immunol Rev 133:93–104

Kisielow P, von Boehmer H 1995 Development and selection of T cells: facts and puzzles. Adv Immunol 58:87–209

Schönrich G, Kalinke U, Momburg F et al 1991 Downregulation of T cell receptors on self-reactive T cells as a novel mechanism for extrathymic tolerance induction. Cell 65:293–304

Schönrich G, Momburg F, Malissen M et al 1992 Distinct mechanisms of extrathymic T cell tolerance due to differential expression of self antigen. Int Immunol 4:581–590

Schwartz RH 1996 Models of T cell anergy: is there a common molecular mechanism? J Exp Med 184:1–8

Steinmann L 1996 A few autoreactive cells in an autoimmune infiltrate control a vast population of nonspecific cells: a tale of smart bombs and the infantry. Proc Natl Acad Sci USA 93: 2253–2256

van Oosten M, van de Bilt E, de Vries HE, van Berkel TJC, Kuiper J 1995 Vascular adhesion molecule-1 and intercellular adhesion molecule-1 expression on rat liver cells after lipopolysaccharide administration *in vivo*. Hepatology 22:1538–1546
Wekerle H, Bradl M, Linington C, Kaab G, Kojima K 1996 The shaping of the brain-specific T lymphocyte repertoire in the thymus. Immunol Rev 149:231–243
Yednock TA, Cannon C, Fritz LC, Sanchez-Madrid F, Steinman L, Karin N 1992 Prevention of experimental autoimmune encephalomyelitis by antibodies against $\alpha4\beta1$ integrin. Nature 356:63–66

DISCUSSION

Allison: Have you looked for other types of infiltrating cells such as APCs? These might also be playing a role: the critical factor may not just be accessibility for T cells.

Hämmerling: Why should we look at infiltration with APCs? The sinusoids are loaded with Kupffer cells anyway.

Allison: What I'm getting at is might there be some initiation of co-stimulation.

Hämmerling: Presumably you are referring to the possibility that we have activated T cells in the liver, but they need another stimulus in order to differentiate into effector CTLs and do something to the hepatocytes. We have thought about this but we don't think it is a likely explanation.

Shevach: You are conditioning the T cell rather than the organ.

Hämmerling: As I said, I don't think this is likely. The CD8 cells are clearly activated: they express all activation markers and are almost certainly functional CTLs. I am not aware of any data showing that activated $CD8^+$ T cells still need another kick before they can cause damage to a target cell.

Allison: The key question concerns how the T cell is activated. The fact that a T cell has been activated previously doesn't mean it is still capable of effector function.

Hämmerling: In control studies we have shown that when we first activate the T cells by breaking tolerance and then inject the tumour, this results in tumour rejection.

Lenardo: We did the experiments with anergy and IL-2 reversal, and there's a trivial explanation which is that most T cell clones already have IL-2 receptor on the surface (Beverly et al 1992). The critical thing is to stimulate them and get them to cycle, and that breaks tolerance. The reason you have a requirement for the additional exposure to antigen is that normal cells probably don't have IL-2 receptor on the surface without encountering antigen.

Hämmerling: The reason I mentioned this is solely to point out that IL-2 alone *in vivo* doesn't seem to break tolerance and, therefore, elevated IL-2 levels do not seem to be terribly dangerous.

Lenardo: No, but you could probably replicate that same effect by taking your cells out after the initial antigen exposure and treating them *in vitro* with IL-2.

Hämmerling: These cells behave similarly to Herman Waldmann's tolerant T cells: *in vitro* they lose their tolerance immediately. That is why we have to do all experiments *in vivo*.

Lenardo: Your adhesion molecule explanation is probably an excellent one for why you don't get the liver damage with the activated cells. We have looked at Fas killing of B cells by T cells, and we were fascinated to find that it's absolutely dependent on adhesion molecules (Wang & Lenardo 1997). If you block adhesion molecules you can have the activated T cells with loads of FasL exposed to activated B cells that have lots of Fas on their surface, and they don't kill. So if CD95 or even the older literature on perforin-dependent killing is relevant to your deletion function, the conditioning may be exactly as you suggest.

Hafler: The obvious next series of experiments seems to be to use anti-B7 or anti-VCAM: have you done any of those experiments?

Hämmerling: Like everyone else who works with transgenic mice, I encounter an everlasting shortage of supply of double-transgenic mice. All these experiments are planned.

Hafler: You call this autoimmunity, which is fair enough, but is there any real precedent for CD8-mediated autoimmune response in terms of animal models?

Hämmerling: In mice, one can induce experimental liver autoimmunity by injecting liver homogenate.

Hafler: Is that CD4 or CD8 mediated?

Hämmerling: This is not clear, Interestingly, in this context, Dr P. Galle in Heidelberg told me that when they perform operations on the liver, for instance removing the gall bladder from patients, they sometimes see autoimmune-like phenomena, which they believe is because they have damaged liver tissue resulting in the release of liver self antigens which will then activate autoreactive, namely liver-specific, T cells.

Hafler: It's a central question in immune regulation. We have been looking hard for autoreactive CD8 cells in humans to MBP and proteolipid protein (PLP) epitopes. Clearly, you can induce CD4-mediated experimental autoimmune disease where you get tissue destruction: can you get by injection of antigen in some fashion, a CD8-mediated autoimmune response? Or are we just looking at a situation similar to diabetes?

Hämmerling: In the field of diabetes it is argued not only that CD4 cells are important, but also that CD8 cells are involved. Our experimental system is a minimalistic approach where we have tried to reduce the reactive cells down to one type. This does not exclude the possibility that other cell types might also be very important. We observe autoimmunity with these clonotypic CD8 cells alone. Actually, the major message of this work is that two different steps are required for the induction of liver-specific autoaggression, namely breakdown of tolerance and conditioning of the target organ.

Lechler: You've discussed a fair bit the importance of quantitative variation in tolerance induction. Every time I see the CRP.K^b data I'm amazed, because you have K^b that is undetectable yet it induces a profound state of tolerance. I would be interested in your comments on mechanistically how that might fit in. Following on from that, have you done the breaking tolerance experiments in the CRP.K^b mice? Is there enough K^b there to cause liver damage if you do break tolerance with the IL-2 tumour? Your data show that there is enough K^b to induce tolerance in these mice: is there enough K^b there to act as an antigen to cause liver damage if you break tolerance in the CRP.K^b mice with the P815.K^b.IL-2 transfectants?

Hämmerling: Yes, but the problem is that in addition to breaking tolerance, we also have to condition the liver. However, the process of conditioning will also up-regulate K^b on hepatocytes in CRP.K^b mice. Consequently, we don't have mice which are very low in K^b. One might argue that the observed up-regulation of K^b in CRP.K^b mice represents a nice model for autoimmunity, because up-regulation of self-antigen is often observed during an inflammatory reaction. The increased expression of self-antigen may actually result in autoimmunity.

Lechler: So in terms of the induction of tolerance, do you think it is a simple matter that there are days in the life of the mouse when it has a bad day, and so there is a burst of CRP leading to transient high levels of K^b expression?

Hämmerling: It is certainly possible that some of the hepatocytes express a fairly high concentration of the K^b molecule. The fetal liver, as you know, is much more accessible for lymphocytes than the adult liver—it is a location for haemopoeisis. A small percentage of $K^{b\ high}$ hepatocytes might be sufficient to induce tolerance, but by PCR one wouldn't see the few cells.

Lechler: Perhaps birth is traumatic. If this is the case there could be a transient acute phase response perinatally, favouring tolerance induction in neonatal T cells due to the accompanying up-regulation of K^b under the CRP promoter.

Mitchison: Robert Lechler, your notion of day-to-day variation focuses on triggering. As I understand it, opinion is moving towards variation in regulation being more important.

Wraith: Günter Hämmerling, I was trying to work out whether it is a peculiar feature about the liver that's interesting in your model, or something about your model that's interesting. Despite what you were saying about pertussis, the fact is if you put activated T cells into an EAE-susceptible mouse, there's no problem about the cells getting into the brain and causing EAE even without pertussis. The same is true of diabetes and arthritis. In your system, if it's the liver that's peculiar, then I'm worried, because how do you explain immune surveillance? We know that the immune system can get at the liver and protect against mouse hepatitis virus, for instance. But in your case, even if the immune system is activated, it doesn't get to the liver. So is it the liver or is it the system?

Hämmerling: The cells must have seen the K^b molecule on the hepatocytes—after all, they have been tolerized by them. One possible explanation is that accessibility of the liver parenchyma changes during development: that in neonatal mice it is more readily accessible for T cells than in adults.

Kioussis: I'm concerned about the liver as a target organ. Post-activation, T cells do home to the liver. In your system you're looking at a situation where the liver is also the target. On top of that, you're transferring Rag^- T cells, but the hosts are not Rag^- mice. So there might be other activated cells that are also homing to the liver, therefore aiding your transgenic T cells to do the damage.

Shevach: The transferred cells can also exert protective effects.

Hämmerling: I can try to give a quantitative interpretation: when we break tolerance in these mice, we don't break tolerance in all T cells simultaneously. At the beginning we may have only a few reactivated cells, which go to the liver, but there are not enough of them to cause detectable liver damage. In contrast, in the transfer experiments, a large number of cells will hit the liver simultaneously. This latter situation might be similar to the experiment where we break tolerance and then give *Listeria*. Breaking of tolerance will yield a large amount of autoreactive cells, and with *Listeria* bacteria we will open the door to the liver: that's why we see liver damage. Liver damage may be there all the time, but because of the self-healing properties of the liver it is at too low a level to be detectable. Unfortunately this hypothesis is difficult to test.

Mitchison: Right at the beginning of this meeting we discussed briefly ideas about the nature of immunological neglect. Günter has described an interesting and novel system in which this has been explored. It seems to me there's been a general shift of opinion in the following direction. Since the idea of neglect was introduced by Zinkernagel and Ohashi (Ohashi et al 1991), the emphasis has been on the dialogue between T cells and APCs. Opinion has been moving—even for antigens expressed on islet cells—towards the old-fashioned sequestration idea. When you say 'it makes liver accessible', in a sense you are reviving old ideas about letting T cells into an otherwise sequestered region. It's clear that the system that you have is somehow biased towards detecting sequestration, rather than the dialogue between T cells and APCs, for two reasons: first, it's a CD8 system, so APCs are less critically important and, second, the liver clearly is normally rather inaccessible to T cells. The problem for people trying to think about this in a very general way, is to discount the biases which are introduced in particular experiments and find the general rules. Perhaps there won't be general rules. In some systems APCs will be more important, and in other cases the old idea of sequestration is more important.

Hämmerling: I agree with you. Bernd Arnold and I deliberately chose a CD8 system in order to minimalize the system and to restrict it to one cell type. The other reason for selecting a CD8 T cell system is that working at a cancer centre, I think this work is also relevant for tumour therapy. Presently, much work is being invested in the development of tumour vaccines and activation of T cells. It is not a problem to activate tumour-specific T cells, but if the tumour is shielded off or sequestered—we should keep in mind that not all tumours are infiltrated—then the activated T cells are not able to get into the tumour to eliminate it. Therefore, we may have to re-think our therapeutic approach. Perhaps we have to do something to the tumour in order to make it accessible to T cells. The reverse situation would apply to autoimmune diseases. Here, we shouldn't look only at the T cells and try to block them. As an alternative we should consider interference on the level of the target organ and prevent accessibility by the activated T cells. I think this is the most important message of our results.

Waldmann: In your system the CD4 cells are missing. Many immune responses of course will start with CD4 and CD8 T cells both involved. A comparable model is the kidney-expressing ovalbumin of Jacques Miller, where T cells were not going in and causing damage. I asked him what the consequences of adding small numbers of CD4 cells were, and he said that a small number of CD4 cells made an enormous difference, allowing the CD8s to enter. The question this therefore raises concerns whether or not in a normal situation the CD4 cells might be the cells that did the first stage.

References

Beverly B, Kang SM, Lenardo MJ, Schwartz RH 1992 Reversal of *in vitro* T cell clonal anergy by IL-2 stimulation. Int Immunol 4:661–671

Ohashi PS, Oehen S, Buerki K et al 1991 Ablation of 'tolerance' and induction of diabetes by virus infection in viral antigen transgenic mice. Cell 65:305–317

Wang J, Lenardo MJ 1997 Essential lymphocyte function associated 1 (LFA-1): intercellular adhesion molecule interactions for T cell mediated B cell apoptosis by Fas/APO-1/CD95. J Exp Med 186:1171–1176

Cross-presentation of self antigens to CD8⁺ T cells: the balance between tolerance and autoimmunity

Christian Kurts, William R. Heath, Francis R. Carbone, Hiroshi Kosaka and Jacques F. A. P. Miller

The Walter and Eliza Hall Institute of Medical Research, Post Office, Royal Melbourne Hospital, Victoria 3050, Australia

Abstract. Upon encounter with foreign antigen, tissue-associated antigen-presenting cells (APCs) migrate to draining lymph nodes to prime specific T cells. Using the transgenic RIP-mOVA model, we recently demonstrated that self antigens derived from peripheral tissues are constitutively transported to draining lymph nodes, and can be presented in association with MHC class I molecules by a bone marrow-derived APC population. This form of class I-restricted presentation of exogenous antigen has been referred to as cross-presentation and can induce activation and proliferation of antigen-specific $CD8^+$ T cells. In the absence of $CD4^+$ T cell help, activation of $CD8^+$ T cells is inefficient, and cross-presentation leads to peripheral deletion of autoreactive $CD8^+$ T cells, acting as a mechanism to maintain self-tolerance. If $CD4^+$ T cell help is available, $CD8^+$ T cell responses to self antigens can be rendered immunogenic, leading to autoreactive responses. Whether autoimmunity results from such responses also depends on the tissue location of the antigen. In RIP-mOVA mice, which express the model antigen mOVA (a membrane-bound form of ovalbumin) in the pancreatic β cells and kidney proximal tubules, OVA-specific $CD8^+$ T cells, activated by cross-presentation, infiltrated the pancreas and caused β cell destruction. Interestingly, however, these cells did not infiltrate the kidney, suggesting that proximal tubular cells are to some extent protected from immune destruction. Analysis of the role of antigen concentration indicates that high doses were required for efficient cross-presentation, suggesting that this pathway is directed towards immune responses to high-dose antigens, such as may be present during viral infection.

1998 Immunological tolerance. Wiley, Chichester (Novartis Foundation Symposium 215) p 172–185

The dogma that only antigens derived from intracellularly synthesized proteins are presented in association with MHC-encoded class I molecules to $CD8^+$ T cells (Bevan 1987, Germain 1994) implies that $CD8^+$ T cells can be activated only by a cell that synthesizes the antigen itself. This poses a major problem since it requires that during infections where pathogens avoid professional antigen-presenting cells

(APCs), peripheral tissue cells must activate $CD8^+$ T cells. Activation of naïve T cells is, however, thought to require co-stimulation signals and these are generally not provided by non-professional APCs. Furthermore, since naïve T cells recirculate mainly through lymphatic tissues (Mackay 1993), they would be unlikely to encounter antigen.

The classical cross-priming (Bevan 1976, Gordon et al 1976, Gooding & Edwards 1980, Carbone & Bevan 1990) and cross-tolerance (von Boehmer & Hafen 1986) experiments suggested that certain exogenous antigens, in these cases antigens on allogeneic cells, can gain access to the class I presentation pathway. In addition, exogenous antigens carried by microorganisms (Yewdell et al 1988) or heat shock proteins (Arnold et al 1995) and tumour antigens (Huang et al 1994) were recently shown to be presented in an MHC class I-restricted way *in vivo*. These findings offer an answer to the above dilemma: exogenous antigens derived from pathogens which infect only non-lymphoid tissues can induce a $CD8^+$ T cell response *in vivo*, using a 'cross-presentation' pathway in professional APCs (Rock 1996). This mechanism could also enhance the efficiency of antigen detection by $CD8^+$ T cells. In the absence of cross-presentation, every $CD8^+$ T cell would have to examine every tissue cell for the presence of antigen, thus requiring a very large number of interactions between peripheral tissue cells and $CD8^+$ T cells. The existence of cross-presentation, on the other hand, would limit the cellular interactions to only those involving $CD8^+$ T cells and the cross-presenting professional APCs.

These studies have all focused on foreign antigens introduced into a host. We have used the transgenic RIP-mOVA (rat insulin promoter/membrane-bound ovalbumin) model to investigate whether cross-presentation can also occur for self antigens. RIP-mOVA mice express the model antigen mOVA in pancreatic islets, kidney proximal tubular cells, thymus and testis, but not in bone-marrow-derived cells. In these mice, the model antigen was demonstrated to migrate to lymph nodes draining antigen-expressing tissues, enter the class I presentation pathway of a bone marrow-derived cell population and activate transgenic OVA-specific $CD8^+$ T cells (OT-I cells) (Kurts et al 1996, Miller et al 1997). If, however, exogenous self antigens such as OVA can be 'cross-presented' in a class I-restricted way *in vivo*, autoimmunity would ensue after activation of autoreactive $CD8^+$ T cells that had escaped thymic censorship. How then does the immune system prevent this form of autoimmunity, while at the same time maintaining responsiveness of $CD8^+$ T cells to cross-presented pathogens? Recent data obtained in the RIP-mOVA system offer an answer to this question. Activation by cross-presentation induces an initial expansion followed by subsequent deletion of the activated $CD8^+$ T cells. If enough activated $CD8^+$ T cells are produced during this initial expansion phase, an immunogenic response ensues. The number of activated cells produced depends on their precursor frequency. In the case of a self antigen which is always present in the body, activation of the few specific $CD8^+$ T cells that leave the thymus each day may generate too few cytotoxic T lymphocytes (CTLs) at a time to induce autoimmunity. The activated $CD8^+$ T cells may then be deleted continuously through activation-induced cell death (Webb et al 1990, Moskophidis et al 1993). This would not be the case for a foreign pathogen which suddenly enters the body. Because pathogen-specific $CD8^+$ T cells have not been deleted, they could accumulate to a higher precursor frequency within the $CD8^+$ T cell repertoire. Activation of these cells would produce enough CTLs to induce an immunogenic response.

In the following we will review the observations in the RIP-mOVA model supporting this model, starting with activation of OT-I cells by cross-presentation of exogenous self antigens, describing their fate after activation by this mechanism, and finally examining the factors governing the balance between autoimmunity and tolerance.

Activation of OT-I cells by cross-presentation of self antigens

OT-I cells adoptively transferred into RIP-mOVA mice were activated in the pancreatic and renal lymph nodes which drain OVA-expressing tissues (Kurts et al 1996, Miller et al 1997). Activation occurred as a result of interaction with a bone-marrow-derived APC capable of presenting exogenous OVA in association with MHC class I molecules. This 'cross-presentation' was essential for activation of OT-I cells in the draining lymph nodes, as demonstrated by lack of T cell activation in bm1→RIP-mOVA.B6 bone marrow chimeras. APCs of the bm1 haplotype, in contrast to those of the C57/B6 haplotype (B6), cannot present OVA to OT-I cells. These results are shown in Fig. 1, using the method of labelling OT-I cells with the fluorescent dye CFSE (5[6]-carboxyfluorescein diacetate succinimidyl ester) prior to adoptive transfer. When CFSE-labelled cells divide, the two daughter cells receive half of the original fluorescence. Generally, a cell that has divided *n* times will display a 2^n-fold reduced fluorescence intensity. Therefore, on a FACS histogram, proliferation of adoptively transferred OT-I cells results in separate peaks of divided cells. In bm1→RIP-mOVA.B6 bone marrow chimeras, OT-I cell proliferation was abolished in the renal lymph node (Fig. 1, 2nd row), whereas it was detectable in B6→RIP-mOVA.B6 mice which served as a positive control (Fig. 1, 1st row).

When RIP-mOVA mice were back-crossed to bm1, no activation of OT-I cells occurred after adoptive transfer. Introduction of B6 bone marrow into these RIP-mOVA.bm1 mice restored the activation of OT-I cells in the draining lymph nodes transfer (Table 1, 3rd row), indicating that cross-presentation by a bone marrow-derived APC alone was sufficient to activate OT-I cells and no interaction with the peripheral tissue cells which expressed the antigen was required. CD4 help was not

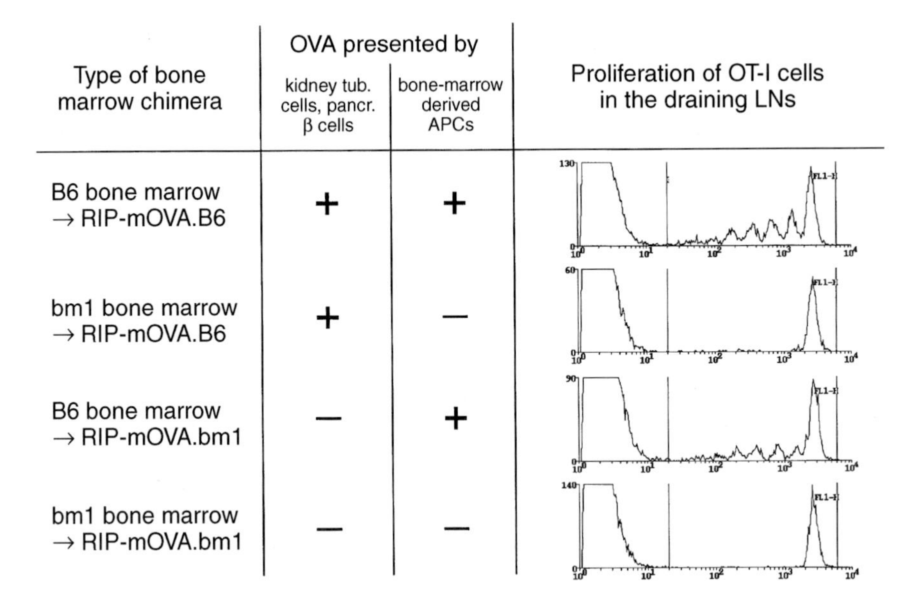

FIG. 1. Activation of OT-I cells by bone marrow-derived cross-presenting APCs. RIP-mOVA mice on a B6 or bm1 background were grafted with bone marrow from either B6 or bm1 non-transgenic mice. Into the four resulting types of bone marrow chimeras 5×10^6 CFSE-labelled OT-I cells were adoptively transferred. Three days later, the renal lymph nodes (LNs) were examined for fluorescent cells by FACS analysis. Histograms were gated for $CD8^+$ T cells. Proliferation of OT-I cells, as evidenced by peaks of divided cells, were seen only when B6 bone marrow was used, regardless of the MHC haplotype of the tissue cells that expressed the antigen.

essential for cross-presentation, because activation of OT-I cells occurred in RIP-mOVA mice depleted of $CD4^+$ T cells by antibody treatment and in RIP-mOVA mice on a $Rag^{-/-}$ background. The latter observation also demonstrates that B cells were not required for cross-presentation.

Aberrant expression of OVA in bone marrow-derived cells was not involved in OT-I cell activation in the RIP-mOVA model, as evidenced by lack of activation of the T cells in non-draining lymph nodes. Another result showing lack of transgene expression in bone marrow-derived cells came from an experiment where bone marrow from RIP-mOVA × OT-I double-transgenic mice was introduced into irradiated non-transgenic B6 mice. OT-I cells matured normally in the thymus of these bone marrow chimeras (unpublished observations). If aberrant expression had occurred in bone marrow-derived thymic APCs, it would have induced negative selection. Furthermore, the bone marrow-derived APCs in the

TABLE 1 Cross-presentation of self antigen leads to deletion of autoreactive CD8$^+$ T cells

No. of OT-I cells transferred	*No. of days after transfer analysis took place*	*OT-I cells recovered from B6→RIP-mOVA.bm1 mice*	*OT-I cells recovered from B6→non-transgenic.bm1 mice*
5×10^6	21	497×10^3 (n=1)	632 ± 10^3 (n=2)
5×10^6	37	$36 \pm 35 \times 10^3$ (n=4)	$485 \pm 240 \times 10^3$ (n=3)
6×10^6	55	$46 \pm 35 \times 10^3$ (n=4)	$423 \pm 118 \times 10^3$ (n=4)

The table summarizes the results of three separate experiments examining the survival of OT-I cells in RIP-mOVA.bm1 mice grafted with B6 bone-marrow, in which only bone marrow-derived cross-presenting APCs can present OVA to OT-I cells. At different time points after adoptive transfer the number of OT-I cells remaining in the lymphatic tissues of the recipient mice was determined by FACS analysis.

periphery of these mice did not activate OT-I cells, again suggesting that they were not expressing the transgene.

The finding that only bone marrow-derived APCs activated OT-I cells in RIP-mOVA mice was exploited to investigate the molecular requirements for cross-priming. Introduction of bone marrow from mice in which the gene encoding the transporter associated with antigen processing (TAP) had been disrupted, abolished cross-presentation. This suggests that exogenous molecules utilized the TAP for loading onto class I molecules of the cross-presenting APCs (unpublished observations). Discovery of the identity of the cross-presenting APC using this approach, however, will have to await the availability of mice lacking defined APC populations.

Cross-priming is not restricted to membrane-bound OVA. It was also observed in RIP-OVA^{hi} mice expressing high levels of soluble OVA without the transmembrane domain that directs OVA to the cell membrane in RIP-mOVA mice, and in RIP-HA mice (unpublished observations), expressing influenza hemagglutinin (HA) in pancreatic islets (Morgan et al 1996). However, in RIP-mOVA^{lo} mice expressing low levels of soluble OVA in the islets (Blanas et al 1996), only very few OT-I cells were activated in the draining lymph nodes. Therefore, the concentration of exogenous antigen determines whether it can access the class I cross-presentation pathway. This might explain why cross-presentation has not been reported in mice expressing lymphocytic choriomeningitis virus (LCMV) antigen in their islets (Ohashi et al 1991, Oldstone et al 1991). If the LCMV protein concentration was too low, it would not have been cross-presented in the pancreatic lymph nodes, and consequently it could not have activated LCMV-specific T cells. Assuming that islet cells do not activate naïve $CD8^+$ T cells (either because these do not enter peripheral tissues, or because islet cells do not express the co-stimulator molecules required), this would result in immunological ignorance.

The fate of OT-I cells activated by cross-presentation

After adoptive transfer of OT-I cells into unirradiated RIP-mOVA mice, activation by cross-presentation can first be detected after 25 h. From then on, every cell cycle requires approximately 4.5 h (Kurts et al 1997). When 5×10^6 OT-I cells were transferred, autoimmune diabetes was induced after 6–8 d. When the number of OT-I cells in the lymphatic tissues was determined at this time, twice as many OT-I cells were found in RIP-mOVA mice as in non-transgenic littermates. Quantitation of OT-I cells three weeks after adoptive transfer, however, revealed that most of these had disappeared from transgenic, but not from non-transgenic mice. This suggests that OT-I cells, activated by cross-presentation, underwent activation-induced cell death after an initial wave of expansion. To investigate whether deletion of these cells was due only to cross-presentation or also to subsequent encounter with antigen on peripheral tissue cells, we transferred OT-I cells into B6→RIP-mOVA.bm1 bone marrow chimeras. As described above, OT-I cells were activated in these mice by cross-presenting APCs derived from the B6 bone marrow, but they could not interact with antigen on peripheral tissue cells. Five to seven weeks after transfer, far fewer adoptively transferred cells remained in B6→RIP-mOVA.bm1 mice, as compared with B6→non-transgenic.bm1 mice (Table 1). Consequently, cross-presentation of self antigens was sufficient to remove autoreactive $CD8^+$ T cells, presumably by deletion.

The adoptive transfer of 5×10^6 OT-I cells is unlikely to mimic the normal situation, where small numbers of newly-matured cells enter the periphery from the thymus each day. We reasoned that diabetes may have occurred because the normal tolerogenic mechanisms could not cope with such a large number of injected T cells. To create a more physiological situation, in which OT-I cells would be generated continuously in the thymus, we manipulated the RIP-mOVA mice in a way that prevented OVA from being expressed intrathymically. Thymectomized RIP-mOVA mice were grafted with OT-I bone marrow and a non-transgenic syngeneic thymus. In the thymus grafts of these 'TG-RIP' mice, OT-I cells matured to an extent similar to that in TG non-transgenic mice. In the peripheral lymphatic tissues, however, five times fewer OT-I cells remained, again demonstrating peripheral deletion of these cells (Kurts et al 1997). When the thymus graft was removed from TG-RIP mice, the few remaining OT-I cells disappeared from the lymphatic tissues. A high proportion of OT-I cells of TG-RIP mice displayed an activated phenotype. This was higher in the pancreatic and renal lymph nodes than in other lymphatic tissues, so that activation in TG-RIP mice presumably took place in these locations as it did after adoptive transfer, and therefore by cross-presentation. When re-stimulated with antigen *in vitro*, the few remaining OT-I cells from TG-RIP mice responded faster than those from RIP non-transgenic mice, indicating that they were not anergized. Although the pancreatic islets in TG-RIP mice were infiltrated by $CD8^+$ T cells, diabetes rarely ensued (1 out of 12 mice after 4 months).

The balance between autoimmunity and tolerance

The lack of diabetes in TG-RIP mice suggested that transfer of a small number of OT-I cells would not result in autoimmunity. Indeed, transfer of 0.25×10^6 OT-I cells did not induce diabetes in 25 recipients, although this number was still clearly in excess of the normal precursor frequency of T cells specific for a given antigen in a normal repertoire. Because of their curtailed lifespan following activation by cross-presentation, these cells might have disappeared before they could destroy sufficient antigen-expressing pancreatic β cells. Titration of the number of transferred OT-I cells showed a clear correlation with the incidence of diabetes, the disease being induced in 50% of the recipients following transfer of 1×10^6 OT-I cells.

Both the $CD4^+$ and the $CD8^+$ T cells in RIP-mOVA mice were specifically tolerant to OVA (unpublished observations). This was most likely due to ectopic expression of OVA in the thymus of RIP-mOVA mice, as demonstrated by thymic deletion of OT-I cells in RIP-mOVA × OT-I, and of OT-II cells in RIP-mOVA × OT-II mice (OT-II cells are transgenic OVA-specific $CD4^+$ T cells, unpublished results). Consequently, adoptive transfer of OT-I cells into RIP-mOVA mice represents a model system in which the response of autoreactive $CD8^+$ cells that have escaped thymic negative selection can be studied. Thus, a very high frequency of OT-I cells, much higher than the average $CD8^+$ T cell frequency for a given antigen in a normal mouse repertoire, was required to induce an immunogenic response in the absence of CD4 help.

When CD4 help was made available by coinjecting OT-II cells, diabetes was induced in recipients of 0.25×10^6 OT-I cells, which alone did not result in disease. OT-II cells injected alone, even in very large numbers, were also not diabetogenic. CD4 help was therefore another factor influencing the outcome of this immune response. Whether OT-II cells affected the response of OT-I cells by increasing their proliferation, prolonging their life-span (Kirberg et al 1993) or rendering them more aggressive is currently being investigated. Recent observations indicate that cognate interaction of the $CD4^+$ and the $CD8^+$ T cells with the same APC is required to induce CTLs by cross-priming (Bennett et al 1997). In the RIP-mOVA model, this would imply that OT-II cells interact with the cross-presenting APC that activates OT-I cells, possibly changing its phenotype to make it more stimulatory to $CD8^+$ T cells.

Importantly, although OT-I cells could induce autoimmune diabetes, they did not infiltrate the kidney (unpublished results). This was not due to the lack of MHC-restricted antigen expression, since purified proximal tubules could stimulate OT-I cell proliferation *in vitro*. The mechanism which excludes activated OT-I cells from the kidney tubules remains to be elucidated.

In the absence of cross-presentation, neither activation of OT-I cells nor their infiltration into pancreatic islets took place. This was shown by injecting naïve OT-I cells into bm1→RIPmOVA.B6 bone marrow chimeras. Two possible explanations could account for this result. Either the cells could not enter islets as long as they were not activated, or they entered in small numbers, but were not activated by antigen-expressing islet cells, e.g. because of lack of co-stimulation. This situation resembles that observed in the RIP-K^b system, in which the K^b-specific $CD8^+$ T cells could recognize antigen only on islet cells, but not in the pancreatic lymph nodes, because the intact class I molecule K^b is not cross-presented. In this case, the specific $CD8^+$ T cells ignored the islets, unless activated by priming with K^b-bearing skin grafts (Heath et al 1995).

Conclusions

The existence of the 'cross-presentation pathway' enables $CD8^+$ T cells to be activated in lymphatic tissues draining the sites of antigen expression. Not only does this represent an efficient mechanism by which $CD8^+$ T cells can scan non-lymphoid tissues for intracellular pathogens without the need to leave the lymphatic compartment, but it also provides a means by which these antigens can be presented in the context of the appropriate co-stimulator molecules.

Our results indicate that cross-presentation of self antigens can activate autoreactive $CD8^+$ T cells that have escaped thymic negative selection. This, however, does not lead to autoimmunity under normal circumstances, because in the absence of $CD4^+$ help these activated $CD8^+$ T cells are subjected to normal homeostatic mechanisms which lead to their deletion.

Acknowledgements

We wish to thank the Walter and Eliza Hall Institute for Medical Research for creating a supportive and inspiring atmosphere. In particular we are grateful for technical assistance by Tatiana Banjanin, Merryn Ekberg, Jenny Falso, Freda Karamelis, Maria Karvelas and Paula Nathan. Janette Allison, Megan Barnden, Paul Gleeson and Leonie Malcolm were involved in the generation of the transgenic mouse lines. The work was funded by grants from the Deutsche Forschungsgemeinschaft, the NIH, the National Health and Medical Research Council of Australia and the Australian Research Council.

References

Arnold D, Faath S, Rammensee H, Schild H 1995 Cross-priming of minor histocompatibility antigen-specific cytotoxic T cells upon immunization with the heat shock protein gp96. J Exp Med 182:885–889

Bennett SR, Carbone FR, Karamalis F, Miller JFAP, Heath WR 1997 Induction of a $CD8^+$ cytotoxic T lymphocyte response by cross-priming requires cognate $CD4^+$ T cell help. J Exp Med 186:65–70

Bevan MJ 1976 Cross-priming for a secondary cytotoxic response to minor H antigens with H-2 congenic cells which do not cross-react in the cytotoxic assay. J Exp Med 143:1283–1288

Bevan MJ 1987 Antigen recognition. Class discrimination in the world of immunology. Nature 325:192–194

Blanas E, Carbone FR, Allison J, Miller JFAP, Heath WR 1996 Induction of autoimmune diabetes by oral administration of autoantigen. Science 274:1707–1709

Carbone FR, Bevan MJ 1990 Class I-restricted processing and presentation of exogenous cell-associated antigen *in vivo*. J Exp Med 171:377–387

Germain RN 1994 MHC-dependent antigen processing and peptide presentation: providing ligands for T lymphocyte activation. Cell 76:287–299

Gooding LR, Edwards CB 1980 H-2 antigen requirements in the *in vitro* induction of SV40-specific cytotoxic T lymphocytes. J Immunol 124:1258–1262

Gordon RD, Mathieson BJ, Samelson LE, Boyse EA, Simpson E 1976 The effect of allogeneic presensitization on H–Y graft survival and *in vitro* cell-mediated responses to H–Y antigen. J Exp Med 144:810–820

Heath WR, Karamalis F, Donoghue J, Miller JF 1995 Autoimmunity caused by ignorant $CD8^+$ T cells is transient and depends on avidity. J Immunol 155:2339–2349

Huang AY, Golumbek P, Ahmadzadeh M, Jaffee E, Pardoll D, Levitsky H 1994 Role of bone marrow-derived cells in presenting MHC class I-restricted tumor antigens. Science 264:961–965

Kirberg J, Bruno L, von Boehmer H 1993 $CD4^+8$-help prevents rapid deletion of $CD8^+$ cells after a transient response to antigen. Eur J Immunol 23:1963–1967

Kurts C, Heath WR, Carbone FR, Allison J, Miller JFAP, Kosaka H 1996 Constitutive class I-restricted exogenous presentation of self antigens *in vivo*. J Exp Med 184:923930

Kurts C, Kosaka H, Carbone FR, Miller JFAP, Heath WR 1997 Class I-restricted crosspresentation of exogenous self antigens leads to deletion of autoreactive $CD8^+$ T cells. J Exp Med 186:239–245

Mackay CR 1993 Homing of naive, memory and effector lymphocytes. Curr Opinion Immunol 5:423–427

Miller JFAP, Heath WR, Allison J et al 1997 T cell tolerance and autoimmunity. In: The molecular basis of cellular defence mechanisms. Wiley, Chichester (Ciba Found Symp 204), p 159–168

Morgan DJ, Liblau R, Scott B et al 1996 $CD8^+$ T cell-mediated spontaneous diabetes in neonatal mice. J Immunol 157:978–983

Moskophidis D, Lechner F, Pircher H, Zinkernagel RM 1993 Virus persistence in acutely infected immunocompetent mice by exhaustion of antiviral cytotoxic effector T cells. Nature 362:758–761

Ohashi PS, Oehen S, Buerki K et al 1991 Ablation of 'tolerance' and induction of diabetes by virus infection in viral antigen transgenic mice. Cell 65:305–317

Oldstone MB, Nerenberg M, Southern P, Price J, Lewicki H 1991 Virus infection triggers insulin-dependent diabetes mellitus in a transgenic model: role of anti-self (virus) immune response. Cell 65:319–331

Rock KL 1996 A new foreign policy: MHC class I molecules monitor the outside world. Immunol Today 17:131–137
von Boehmer H, Hafen K 1986 Minor but not major histocompatibility antigens of thymus epithelium tolerize precursors of cytolytic T cells. Nature 320:626–628
Webb S, Morris C, Sprent J 1990 Extrathymic tolerance of mature T cells: clonal elimination as a consequence of immunity. Cell 63:1249–1256
Yewdell JW, Bennink JR, Hosaka Y 1988 Cells process exogenous proteins for recognition by cytotoxic T lymphocytes. Science 239:637–640

DISCUSSION

Allison: Do you have any speculation as to why the CD8-only system sees deletion? Granted, you can't renew the cells, but you do have a virtually endless supply of antigen and APCs. In the absence of help, why do they go away?

Kurts: Your question concerns whether the observations of autoimmunity after introduction of CD4 help and deletion in the absence of CD4 help are linked to each other. Indeed, they are: introduction of CD4 help also impairs the deletion of $CD8^+$ T cells, although this result is preliminary at present. Therefore, $CD4^+$ T cell help seems to play a major regulatory role for the immune response induced by cross-presentation of self antigen.

Allison: I could offer another explanation based on some *in vitro* work. We gave primary $CD8^+$ T cells a big co-stimulatory signal sufficient for them to make enough IL-2 to proliferate and differentiate to effector function. However, they began to lose that ability. This result is fairly counterintuitive, but this is what the data were telling us. They were getting more and not less dependent on exogenous IL-2.

Kamradt: Do you have any idea what the difference between the kidney and the pancreas might be? Why do the T cells attack the mOVA-expressing cells in the pancreas but not in the kidney?

Kurts: Do you have a suggestion?

Kamradt: A trivial suggestion could be that the level of expression of co-stimulatory molecules differs between the two organs.

Kurts: This was also our first idea. To address this question we cultured OT-I cells with purified kidney tubular cells and they proliferated vigorously. This demonstrated that tubular cells do express enough antigen to be recognized by OT-I cells.

Mitchison: That sounds like sequestration.

Kurts: That is another possibility. For example, the dense tubular basement membrane that surrounds the tubules might be impenetrable to $CD8^+$ T cells.

Stockinger: What worried me when I heard these data was that I used to believe what Polly Matzinger said: that dendritic cells do not move around carrying tissue-specific antigens unless they're being annoyed and activated. In your system it looks as if there is constitutive transfer of antigen from the pancreas into the draining lymph node. Can you exclude the possibility that the expression of these high levels of mOVA bothers the islet cells in some way?

Kurts: This cannot be ruled out completely. However, RIP-mOVA mice never develop diabetic or islet infiltration with leukocytes, even at two years of age. And we have several other lines expressing OVA in the islets and none of these ever become diabetic on their own, unless injected with OT-I cells. Therefore, the transfer of OT-I cells is responsible for inducing diabetes and not the mere presence of the transgene.

Hämmerling: Are these mice specific pathogen-free? That might be relevant.

Kurts: No.

Hämmerling: If they're 'dirty' mice, you might expect to have activated, differentiated dendritic cells around which can take up the antigen. According to Levitsky et al (1994), an antigen expressed by a tumour won't be passed on to the dendritic cells. But if the tumour co-expresses GM-CSF then this will activate the dendritic cells and enable them to take up the antigen and put it into the class I pathway. I could imagine the same thing happening in 'dirty' mice.

Kurts: I have no problems with dirty mice. It reflects the real-life situation, in which we all are constantly exposed to pathogens. No-one lives in a pathogen-free environment.

Stockinger: I have no problems with cross-presentation getting antigen into this pathway. The question is whether peripheral tissue-specific antigen is always present in lymph nodes. This was one of the reasons why one could argue that autoimmunity is not normally induced, because it requires priming in lymph nodes and the antigen isn't there.

Kurts: If antigen does not find its way to the draining lymph node, ignorance of $CD8^+$ T cells will result. On the other hand, if antigen is present but unable to access the cross-presentation pathway, ignorance would also result. Finally, if antigen reaches the cross-presentation pathway, then priming will occur, but the $CD8^+$ T cells will be deleted in the absence of CD4 help.

Jenkins: On the basis of our recent work, antigen presentation in lymphoid tissues should not lead to productive T cell activation as long as inflammation is absent.

Lechler: But going back to Brigitta Stockinger's point, we don't know for sure that there's an absence of inflammation, unless you can reassure us that it is in no way deleterious to the pancreatic β cells for them to be making the levels of mOVA that they are. The possible difference between the kidney and the pancreas is that although the dendritic cells must be trafficking in both cases, the ones from the pancreas may be more activated because there is some low-level β cell damage which is injurious and triggers the dendritic cells.

Kurts: This is possible, but we have no evidence suggesting tissue damage. Of course it is always possible to assume that the presence of the transgene induces differences too subtle to be observed.

Hafler: That's one possibility. Another relates to the difference between CD4 and CD8 cells and their co-stimulation requirements. We've compared CD8 cells with CD4 cells using transgenic and proteolipid protein (PLP) peptides (Höllsberg et al 1997). Whereas our CD4 clones follow the normal rules of no co-stimulation and become anergic or don't re-stimulate unless we add IL-2, the CD8 cells appear to be quite different: even with co-stimulation they become anergic. They become very IL-2 dependent.

Kurts: I have never observed any of these CD8 cells as being anergic.

Hafler: Have you looked at them in mixed culture?

Kurts: Yes.

Allen: We were interested in the kidney proximal tubules because they express class II MHC. Several years ago we showed that they did not express co-stimulatory activity (Hagerty et al 1994). This supports the idea that these cells might be expressing these antigens, but they're really not going to be optimal. So I think Jim Allison's point that they are not going to make IL-2 and will therefore die is relevant.

Kurts: How would you then explain the fact that OT-I cells proliferate in response to tubular cells *in vitro*?

Allen: When you take these cells out, you have to do many things such as enzymic digests, and what may happen *in vitro* is very different from what is occurring *in vivo*.

Kurts: What differences would you expect? Have you observed tubular cells up-regulating MHC class I or co-stimulatory molecules?

Allen: I don't know. We have never directly compared the class II levels *in vivo* and *in vitro*, but I think it could happen.

Stockinger: I don't see why the level of co-stimulation on this secondary organ would be important, if they have already been activated by dendritic cells in the draining lymph nodes. Then they are effector cells which don't need co-stimulation anymore.

Jenkins: Perhaps the dendritic cells that sit in normal lymphoid tissue are not effective T cell stimulators unless they are activated by inflammatory mediators.

Allison: While it's true they can mediate effector function, who knows how many times they can kill? They are certainly not going to make IL-2 and proliferate anymore unless they get co-stimulated again or they get help from CD4s: that was my point. We observed that they seemed to get less responsive with time, even when they've got co-stimulation and were making their own IL-2. This is sort of like the idea that if you have a lot of cells, you can get enough cellular damage to make a difference, but if you don't have enough they just run down.

Lenardo: So the prediction would be that if you put a continuous infusion (Alzet) pump with IL-2 in those mice, you will get more cells and therefore damage.

Mitchison: Is it clear whether the need for bone marrow-derived cells is because they take up the ovalbumin or because, having taken up antigen, they're then better at stimulating? Is it because they have co-stimulatory molecules or is it because they are just better at taking up the ovalbumin?

Kurts: Are you implying that OT-I cells interact with tubular cells *in vivo*, albeit at numbers too low to be observed, but would not become activated due to a deficiency of the tubular cells?

Mitchison: Yes. If kidney cells take up ovalbumin to the same extent as professional APCs, would they then stimulate? That's a difficult question to answer. An easier question is, do you in fact see much more of the ovalbumin on APCs than you do on other non-expressing cells?

Kurts: How would you determine the amount of OVA presented by MHC class I on APCs in the draining lymph node?

Mitchison: By elution.

Kurts: That would require an antibody which recognizes the OVA peptide in association with MHC class I. Angel Porgador from Ron Germain's lab has made such an antibody (Porgador et al 1997). I have tried it, but it did not stain the APCs when I analysed them by flow cytometry.

Wraith: You've shown that cells proliferate in the draining lymph node, but you have no evidence how the antigen gets there. We have heard a lot about dendritic cells: Uwe Staerz was one of many who showed some years ago that macrophages are perfectly good at presenting antigens such as OVA to cytotoxic T cells (Debrick et al 1991). Is the antigen dribbling out? Is it being carried by macrophages or dendritic cells?

Kurts: The phenotype of the cross-presenting APC is still unknown.

Hämmerling: Have you been able to isolate dendritic cells from the draining lymph node and use them to stimulate OT-I cells in a functional assay?

Kurts: We tried that using Ken Shortman's methods to purify out these dendritic cells and there was no response at all.

Hämmerling: So what makes you so certain that priming takes place in the lymph node? Can you formally exclude the possibility that somehow the T cells are primed in the tissue and then they go back to the lymph node?

Kurts: I did not have enough time to show the nephrectomy experiments we published last year, which answer this question. The idea was to remove the left kidney, transfer OT-I cells later and then compare antigen recognition in both renal lymph nodes. This experiment showed that OT-I cells still recognized antigen in the kidney lymph node when the kidney, which is the antigen source, was removed shortly before transfer of the T cells. However, when they were transferred 10 or more days after nephrectomy, antigen recognition was abrogated. Therefore, antigen is derived from the kidney, T cells see it in the draining lymph node, and it has a limited half-life.

Cornall: Along those lines, have you looked at earlier time points for CD69?

Kurts: Yes, they express CD69 very early.

Mitchison: Could I make a very brief point about your evidence against transfer of whole class I molecules. Your experiment is one of several which argue strongly against the transfer of whole MHC molecules with or without their peptides attached. Nevertheless, it is true that cell surface glycoproteins are co-processed on the same cell membrane in the immune system when they act as allo antigens (Fisher et al 1989). You can get help from one minor antigen to another minor antigens, or a minor antigen can serve as a helper epitope for an MHC molecule. Thus there is a certain amount of transfer of glycoproteins, including MHC molecules, from one cell to another under those conditions. It's quite evident that it doesn't show up there, but I would just like to mention the fact that it is possible and does happen sometimes.

Hämmerling: I wonder how likely this possibility of MHC transfer is. If an antigen is transferred from one cell to another, then it usually goes through the endocytic pathway where it is processed, i.e. degraded. I don't know of any example where a complete MHC molecule is transferred and then stuck into the membrane of the other cell in a functionally active form.

References

Debrick JE, Campbell PA, Staerz UD 1991 Macrophages as accessory cells for class I MHC-restricted immune responses. J Immunol 147:2846–2851

Fisher AG, Goff LK, Lightstone L et al 1989 Problems in the physiology of class I and class II MHC molecules, and of CD45. Cold Spring Harb Symp Quant Biol 54:667–674

Hagerty DT, Evavold BD, Allen PD 1994 Regulation of the co-stimulator B7, not class II MHC, restricts the ability of murine kidney tubule cells to stimulate $CD4^+$ T cells. J Clin Invest 93:1208–1215

Höllsberg P, Scholz C, Anderson DE et al 1997 Expression of a hypoglycosylated form of CD86 (B7-2) on human T cells with altered binding properties to CD28 and CTLA-4. J Immunol 15:4799–4805

Levitsky HI, Lazenby A, Hayashi RA, Pardoll DM 1994 *In vivo* priming of two distinct antitumor effector populations: the role of MHC class I expression. J Exp Med 179:1215–1224

Porgador A, Yewdell JW, Deng YP, Bennick JR, Germain RN 1997 Localization, quantitation and *in situ* detection of specific peptide MHC class I complexes using a monoclonal antibody. Immunity 6:715–726

General discussion III

Stockinger: What do you think about the idea that there is constitutive transfer of tissue antigens via dendritic cells (DCs) into the draining lymph nodes? Presumably this wipes out the whole idea of sequestration.

Kurts: We do not believe that all antigens are cross-presented. In preliminary experiments we have found that a certain antigen dose must be overcome to allow antigen to be cross-presented. Antigens expressed at a lower concentration would fail to enter this pathway and consequently will be ignored. This might explain why in some models cross-presentation was not reported (Ohashi et al 1991, Oldstone et al 1991).

Abbas: I'm not sure that any of us can or would be willing to tackle the question as to whether this happens with all self antigens. But one of the points I thought I was trying to make on the first day was that if you accept that getting rid of Fas ligand or CTLA-4 leads to autoimmunity, then you have to accept that recognition of self is a normal phenomenon. It is absolutely inescapable. You've not changed signal 2, or danger, or anything else — you have just taken away two proteins that regulate what happens to T cells. This implies that recognition of self is a normal phenomenon, and the guess is it's happening in draining lymph nodes. I don't see how to get around that argument.

Stockinger: I'm not sure that I accept that this is autoimmunity.

Abbas: That's a much tougher problem.

Hafler: Perhaps one of the fundamental questions you can address is as follows: is autoimmunity a fundamental physiological process to regulate tissue damage? Therefore there is a logical loading of certain APCs and there's normal T cell recognition of antigen where there is tissue destruction, not necessarily leading to further tissue destruction but to scar formation.

Shevach: In my paper (Shevach et al 1998, this volume) I describe a situation where regulatory T cells play a clear role in protecting against the development of autoimmunity. In the model we have studied, the autoantigen is the enzyme H/K ATPase, expressed by the gastric parietal cell. It is an abundant protein produced by a cell that turns over rapidly. We believe that some of this antigen drains to the gastric lymph node, is processed by DCs present in the node, and is always expressed on the surface on these cells in association with MHC class II molecules. It is capable of being recognized by autoreactive T cells which migrate through the node, yet disease is not seen because of the protective effects of regulatory T cells.

Mitchison: Let's now ask the question, what about proteins which are less abundant? The one that comes to my mind is allo-HPPD, or F liver protein as it used to be called. This is an abundant protein in the liver and is clearly passively acquired by the thymus (Wedderburn et al 1984).

Stockinger: But you showed it's in the blood.

Mitchison: I was just going to add that. Would a fair generalization be that any self protein which gets transferred to antigen-presenting cells (APCs) is likely to be detectable by ELISA in serum? For example, if you look in the fluid draining the stomach, do ELISAs pick up your protein?

Shevach: It would be technically very difficult to isolate gastric lymph and assay for the presence of a protein in that fluid. Of course, the antigen we have identified is a peptide fragment of the enzyme and it is not yet known whether this is an immunodominant epitope.

Mitchison: There's not going to be any simple answer. Some proteins circulate, sometimes peptides derived from abundant proteins get passed down in the lymph, probably there are many examples of proteins which never get out and never get anywhere.

Stockinger: That's a question we always fail to address. There are all these assumptions that massive tissue destruction results in antigens getting into the circulation, maybe inducing subsequent tolerance in the thymus, and so on. Nobody has managed to show that. There was a situation—one of Diane Mathis' transgenics where the whole pancreas was destroyed—which would have been an ideal situation to get the antigen out there, but there was no influence on the thymus (Katz et al 1993). I think we are always waving this question aside.

Allen: Just to follow up on that, I think that the immune system is so exquisitely sensitive that we are often faced with a problem of detection in our experiments. We have functional read-outs and want a biochemical basis. We're still at a level where things are happening biologically that we can't detect biochemically.

Mitchison: There are two other issues which I would like us to discuss. The first is the relative importance of triggering versus regulation. The second is the role of adjuvants or inflammatory cytokines in raising protective T cells, and whether that's a reasonable therapeutic option. Let's start with the first one: triggering versus regulation: I know that's been very much on your mind, Thomas. Would you like to say a few words about it?

Kamradt: I can reference some data which made us think about this. This is Pam Ohashi's model. She used T cell receptor (TCR) transgenic mice. The TCR was specific for the lymphocytic choriomeningitis virus (LCMV) glycoprotein. She showed that those T cells could cross-react with a peptide from an enzyme, dopamine β-monooxygenase, that is expressed in the adrenal medulla. When the transgenic mice were infected with LCMV she showed that some T cells migrated into the medulla. There was a slight change in dopamine metabolism but no gross autoimmunity. The question arises as to what regulatory mechanisms are preventing severe autoimmunity in this model.

Kioussis: Were these transferred cells activated?

Kamradt: Yes, they were specific for a LCMV glycoprotein. Some part of the explanation obviously relates to numbers: in normal BL/6 mice there were no signs of pathology at all. Since the LCMV glycoprotein-specific TCR derived from BL/6 mice, it is clear that they have this potentially cross-reactive receptor in their repertoire. However, only in TCR transgenic mice were there enough LCMV glycoprotein-specific T cells to be detected in the adrenal medulla.

Hafler: In terms of cross-reactive antigens, it's clear that some cross-reactive peptides can induce Th2-type responses. Getting back to the point that Abul made earlier about Kim Bottomly's work, we in fact have similar data to Kim in terms of cross-reactive peptides that switch cytokine secretion. So a peptide that's cross-reactive with self antigen but provides a weaker signal may well be switching cytokine secretion which may then affect whether one can drive autoreactivity to effect autoimmunity and tissue destruction. We have to differentiate between the idea of self-recognition and tissue destruction: they are clearly two different events.

Mason: This is just a hypothesis. Cytotoxic T cells have evolved to kill virus-infected cells. If a cell is not virus-infected there's no real reason why it should be killed. Is there any evidence that a cell that is virus-infected is actually putting out molecules which help the cytotoxic T cell kill it? Is this perhaps the explanation for these negative results in tissue which is otherwise healthy?

Mitchison: I would like to comment on that question, and throw in a piece of information. It's the view of Jonathan Howard that a virus-infected cell turns into an antigen-presenting machine because it has no other function. An important signal for it to start that developmental phase is of course γ-interferon, which up-regulates class I expression. It turns out that γ-interferon induces a great deal more than that, and Howard's group in Köln are engaged in a molecular-biological dissection of what goes on. I think you're right, Don: there is a whole change in phenotype which is in the process of being defined.

Wraith: Any cell can become a target for an activated CD8 killer cell, but that's not to say that any virus-infected cell will prime a naïve CD8 cell.

Mitchison: That's true, but what I'm saying is that if you look at what γ-interferon induces in a fibroblast, it's like the differentiation of a new cell type.

Cornall: With reference to Gunter Hämmerlings's results and Thomas Kamradt's comments, Srinivas Akkaraju, a graduate student in Chris Goodnow's lab, has made transgenic mice expressing HEL under either thyroglobulin or insulin promoters (Akkaraju et al 1997). He has crossed these mice with 3A9 TCR transgenic mice specific for HEL. The TCR transgenic T cells see antigen and are functionally less reactive, although they can be activated *in vitro*. All the mice have insulitis or thyroiditis, but do not progress to overt autoimmune disease. Presumably, partial activation of the T cells is normally sufficient to prevent disease, which might be initiated by some additional insult. Interestingly, and as an aside, B cells do not see the tissue-bound antigens.

Kamradt: Does that infiltration go away?

Cornall: I think it must persist because it is in the tissue of mice at all ages examined.

Kamradt: So it persists but they never get sick?

Cornall: Yes. They retain enough normal tissue to do well.

Mitchison: There is a substantial literature on insults to organs resulting in transient autoantibody production. As far as I know, much less is known about the transient T cell activation which presumably induces the antibodies, but if frequent triggering of T cells occurs, and if only one in a million of these insults ever ends up in chronic autoimmunity, the controls must be enormously efficient.

Allen: The best numbers are in heart attacks. Millions of people world-wide have ischaemic heart disease, and the incidence of any kind of autoimmune disease resulting from that is incredibly rare.

Hafler: We have compared stroke tissue with multiple sclerosis (MS) tissue in terms of co-stimulatory molecules, even in the same brain where people with MS have died from stroke (Windhagen et al 1995). In the MS plaque there is B7-1 and B7-2 expression, whereas in stroke there are sheets of T cells with only B7-2. I don't know the relative roles of these two co-stimulatory molecules, but this suggests that the different co-stimulatory molecules that are present in tissue damage will dictate whether one gets scar formation or frank autoimmune disease.

Mitchison: That's a pretty clear hypothesis: B7 is extremely dangerous. If you express it in tissue along with tissue damage you're in deep trouble. I can hardly believe it's as simple as that.

Cornall: The flip side to a lot of this discussion is the fact that a lot of autopsy reports quite frequently show thyroiditis, without any history of overt disease.

Mitchison: And do you think those are all due to B7 having put in an appearance?

Shevach: B7 and a sprinkling of IL-12 will do it.

Hafler: But that's actually what we found: MS plaques have IL-12 and B7-1 and in stoke tissue there's no IL-12.

Lenardo: One exception is sympathetic ophthalmia, where you have damage to one eye and then at a later time the other eye becomes transiently inflamed and can suffer tissue destruction. I thought this was T cell mediated. There can be T cell reactions that may be secondary to damage.

Mitchison: That's surely a counter example, because sympathetic ophthalmia is so frequent that in conditions where the opportunities for more sophisticated therapy aren't open — during warfare, for example — it used to be the practice simply to take the eye out.

Wraith: Your original point concerned the difference between activation and regulation. It's quite interesting to think of the various different levels for T cell receptor transgenics specific for myelin basic protein (MBP), because they reveal all the various elements you were considering. If you think of Joan Goverman's mice, here you have mice which are perfectly tolerant of MBP unless they are housed in non-sterile conditions, in which case they get disease. You can take these 'dirty' animals and put them on antibiotics and they don't get disease. You can take our mice, which are transgenic for a similar receptor and they don't get any disease even when housed in non-sterile conditions. Then you can take Charlie Janeway's mice, and they get spontaneous disease, but only if they're on a Rag background. If they have other receptors there is regulation. So there's obviously regulation going on; there's obviously activation by cross-reactivity, but for the most part, unless the cells are activated they don't do any harm.

References

Akkaraju S, Ho WY, Leong D, Canaan K, Davis MM, Goodnow CC 1997 A range of CD4 T cell tolerance: partial inactivation to organ-specific antigen allows nondestructive thyroiditis or insulitis. Immunity 7:255–271

Katz JD, Wang B, Haskins K, Benoist C, Mathis D 1993 Following a diabetogenic T cell from genesis to pathogenesis. Cell 74:1089–1100

Ohashi PS, Oehen S, Buerki K et al 1991 Ablation of 'tolerance' and induction of diabetes by virus infection in viral antigen transgenic mice. Cell 65:305–317

Oldstone MB, Nerenberg M, Southern P, Price J, Lewicki H 1991 Virus infection triggers insulin-dependent diabetes mellitus in a transgenic model: role of anti-self (virus) immune response. Cell 65:319–331

Shevach EM, Thornton A, Suri-Payer E 1998 T lymphocyte mediated control of autoimmunity. In: Immunological tolerance. Wiley, Chichester (Novartis Found Symp 215) p 200–217

Wedderburn L, Lukic ML, Edwards S, Kahan MC, Nardi N, Mitchison NA 1984 Single-step immunosorbent preparation of F-protein from mouse liver with conservation of the allo-antigenic site, and determination of concentration in liver and serum. Mol Immunol 21:979–984

Windhagen A, Newcombe J, Dangond F et al 1995 Expression of co-stimulatory molecules B7-1 (CD80), B7-2 (CD86) and interleukin 12 cytokine in multiple sclerosis lesions. J Exp Med 182:1985–1996

Tolerance induction in mature T lymphocytes

Judith Alferink, Anna Tafuri, Alexandra Klevenz, Günter J. Hämmerling and Bernd Arnold

Division of Molecular Immunology, Tumor Immunology Program, German Cancer Research Center, Im Neuenheimer Feld 280, 69120 Heidelberg, Germany

Abstract. T lymphocytes with self-destructive capacity are often found in healthy individuals, indicating efficient control mechanisms that prevent autoimmunity. Recently, we were able to demonstrate the existence of peripheral tolerance in double-transgenic mice expressing the foreign histocompatibility antigen H-2K^b exclusively outside the thymus and a T cell receptor (Des.TCR) directed against the K^b molecule. In mice expressing K^b only on keratinocytes anti-K^b T cells were still present but failed to reject K^b-positive tissue grafts. This observation would imply a continuous migration of naïve T cells exported from the thymus into non-lymphoid tissues where these fresh thymic emigrants would need to be tolerized. However, this is in contrast to the view that migration to peripheral tissues is restricted to activated T cells. To investigate whether there is a continuous process of tolerization of naïve T cells in adult Des.TCR × 2.4Ker-K^b mice, 2.4Ker-K^b mice were crossed with Rag-2-deficient mice and reconstituted with bone marrow cells of Des.TCR transgenic mice (Des.TCR × 2.4Ker-K^b.Rag-2$^-$). Tolerance was not observed in these chimeric mice. We conclude from these results that in contrast to the neonate the adult physiological environment does not allow tolerance induction to antigens expressed on keratinocytes in T cells newly exported from the thymus. Furthermore, we have to postulate regulatory events responsible for the maintenance of peripheral tolerance in the adult Des.TCR × 2.4Ker-K^b animals.

1998 Immunological tolerance. Wiley, Chichester (Novartis Foundation Symposium 215) p 191–199

Immunological tolerance is acquired during development of the immune system. In the course of thymic maturation newly developing T cells undergo apoptosis upon encountering their specific antigen. However, because not all self antigens are expected to be expressed in the thymus, additional processes operating outside the thymus are required to assure self-protection. A detailed knowledge of such processes is essential for therapeutic intervention either to silence T lymphocytes in autoimmune diseases and in organ transplantation, or to break non-responsiveness to tumour antigens.

The consequences of extrathymic antigen expression on T cell reactivity have been studied using various transgenic mouse models in which non-thymic cells expressed a 'foreign' protein under the control of tissue-specific promoter elements (Miller et al 1992, Arnold et al 1993). In some of these studies the tissue-specific antigen had no effect on the immune system (Zinkernagel et al 1991), whereas in others either deletional (Bertolino et al 1995) or non-deletional tolerance induction was observed (Schönrich et al 1991). The outcome was dependent on the nature of the antigen as well as on the site and amount of expression. For example, in our own studies we showed that expression of the major histocompatibility (MHC) class I antigen K^b in different tissues led to tolerance with distinct alterations in a particular T cell population which carried a K^b-specific, transgenic T cell receptor identified by the anti-clonotypic antibody Désirè-1 (Des.TCR) (Schönrich et al 1992, Ferber et al 1994).

In mice expressing K^b under the glial fibrillary acidic protein promoter only on cell types of neuroectodermal origin, a strong reduction of splenic clonotype$^+$ $CD8^+$ T cells was observed in comparison to the numbers found in Des.TCR single-transgenic mice. This reduction was due to down-regulation of TCR and CD8 molecules. In contrast, significant numbers of clonotype$^+$ $CD8^+$ T cells were present in lymphoid organs of animals with K^b expression only on epithelial cells in skin, tongue and foot pads under the 2.4 kb keratin IV promoter (2.4Ker-K^b). As the Des.TCR$^+$ T cells see the K^b molecule only in its intact form carrying a particular peptide (Tafuri et al 1995), we have to assume that the observed tolerance in these systems is the result of a direct interaction between the T cells and the particular cell type expressing the tissue-specific self (K^b) antigen. This would require a rapid and complete circulation of all T cells, enabling them to meet the K^b antigen on very few cells in the body, such as glial cells in the gut or certain keratinocytes. Studies on migration patterns of T cells, however, do not support this view (Mackay 1991). In the adult sheep, naïve T cells have rarely been found to enter tissues and drain to a regional lymph node. Therefore, naïve T cells would not normally come into contact with autoantigens expressed in non-lymphoid tissues.

To clarify this apparent contradiction, we addressed the following questions. Firstly, is the observed tolerance in (Des.TCR × 2.4Ker-K^b)F1 mice a feature of each individual $CD8^+$ Des.TCR$^+$ T cell or of the total cell population? Secondly, does the expression of antigen on keratinocytes in the adult mouse lead to tolerance induction of $CD8^+$ Des.TCR$^+$ T cells newly exported from the thymus?

$CD8^+$ Des.TCR$^+$ T cells of K^b-tolerant mice are heterogeneous in regard to activation marker expression and lymphokine secretion

Des.TCR × 2.4Ker-K^b mice are tolerant to K^b as judged by acceptance of K^b-positive skin and tumour grafts. Nevertheless, in spleen and lymph nodes of these mice, $CD8^+$ Des.TCR^+ T cells are present in significant numbers. Some of these T cells were found to display enhanced levels of CD2 and CD44 expression in comparison to the CD8+ Des.TCR+ T cells from K^b-reactive Des.TCR transgenic mice. An increase in expression of these proteins has been described for activated T cells (Sprent 1994) indicating that some of the $CD8^+$ Des.TCR^+ T cells had been in contact with the K^b antigen. Expression of the interleukin (IL)-2 receptor remained low and unchanged.

Next, lymphokine production of $CD8^+$ Des.TCR^+ T cells was determined on a single cell level by intracellular staining and three-colour fluorescence analysis after antigen stimulation *in vivo*. For this purpose (Des.TCR $\times$ 2.4Ker-K^b)F1 and Des.TCR mice were injected with syngeneic P815 tumour cells transfected with the K^b gene. B cell-depleted spleen or lymph node cells were analysed 15–20 d later for IL-2, IL-4, IL-10 and interferon (IFN)-γ secretion. IL-2- and IFN-γ-secreting $CD8^+$ Des.TCR^+ T cells were present in equal numbers in the K^b-tolerant double-transgenic and in the K^b-reactive single-transgenic animals. However, a major population of $CD8^+$ Des.TCR^+ T cells secreting enhanced levels of IL-10 and IL-4 was present in the K^b-tolerant mice and absent in the K^b-reactive animals. These observations demonstrate that *in vivo* tolerant cells are not totally inert but can respond to antigenic stimulus, in this case with the production of cytokines that are associated with negative regulation (Groux et al 1996).

In contrast, challenge with K^b-positive tumours that co-express the IL-2 gene resulted in reversal of tolerance. These mice were now able to reject K^b-positive tumour grafts up to 120 days later (see Limmer et al 1998, this volume). Similarly, a strong *in vitro* cytotoxic T lymphocyte (CTL) response was generated when $CD8^+$ Des.TCR^+ cells from (Des.TCR $\times$ 2.4Ker-K^b)F1 mice were stimulated with splenocytes of C57BL/6 mice without addition of exogenous IL-2 (Alferink et al 1995).

Taken together, these findings suggest that tolerance in (Des.TCR $\times$ 2.4Ker-K^b)F1 mice might be based on a balance between naïve and tolerant anti-K^b T cells. Challenge with the K^b antigen *in vivo* leads to strong expansion of IL-4- and IL-10-secreting T cells which might control reactivity of the naïve anti-K^b T cells. However, stimulation with K^b in the presence of IL-2 *in vivo* or under appropriate *in vitro* conditions may result in activation of the naïve anti-K^b T cells and their predominant expansion causing reversal of tolerance.

Continuous presence of antigen is required for maintenance of tolerance

Since the tolerant T cells can easily recover their responsiveness under appropriate stimulation conditions, we investigated whether persistence of antigen was required for maintenance of the tolerant state *in vivo*. We addressed this question by transferring B cell-depleted splenocytes of (Des.TCR $\times$ 2.4Ker-K^b)F1 double-transgenic and Des.TCR single-transgenic mice into low-dose irradiated *nu/nu*

BALB/c mice which do not express the K^b antigen (Alferink et al 1995). To investigate the 'parked' cells for their capacity to reject K^b-positive skin grafts, we grafted the recipients with K^b-positive skin on their flanks 3, 10, 15 or 20 d after transfer. Control mice either received spleen cells from Des.TCR single-transgenic mice or no cells before grafting. Control *nu/nu* mice, which had not received cells, failed to reject the skin graft. As expected, most *nu/nu* mice injected with spleen cells of Des.TCR transgenic mice rejected the grafts. In contrast, when T cells of the tolerant (Des.TCR × 2.4Ker-K^b)F1 transgenic mice were transferred, graft rejection was dependent on the time-span between cell transfer and grafting. Recipients failed to reject skin transplanted 3 d after cell transfer. When the tolerant T cells were parked in BALB/c *nu/nu* mice for a longer period before grafting (15 or 20 d) all recipients rejected the graft. These results demonstrate that tolerant T cells from (Des.TCR × 2.4Ker-K^b)F1 transgenic mice require continuous contact with the K^b antigen to keep their tolerant state *in vivo*. Tolerance wanes within 15 d in the absence of antigen. Since (Des.TCR × 2.4Ker-K^b)F1 mice contain not only T cells carrying the transgenic TCR but also T cells expressing endogenous TCR, it was possible that the recovery of responsiveness after transfer of tolerant populations into *nu/nu* mice was due to expansion of T cells with endogenous TCR so that numbers sufficient for graft rejection were present after 15 d. This is unlikely because then one would expect third-party grafts also to be rejected within 15 or 20 d like the Kb grafts. However, rejection of third-party grafts such as H-2^s skin required at least 60 d. In addition, no expansion of cells with endogenous TCR was observed in the recipients. Therefore, the results strongly indicate that it was indeed the tolerant $CD8^+$ Des^+ T cell population which acquired responsiveness in the antigen-free environment of the *nu/nu* BALB/c mice.

Naïve T cells are not tolerized by keratinocyte-specific antigens in the adult animal

To investigate whether there is a continuous process of tolerization of naïve T cells by encounter with the K^b antigen on keratinocytes of adult (Des.TCR × 2.4Ker-K^b)F1 mice, we crossed 2.4Ker-K^b mice onto a Rag-2-deficient background and reconstituted the progeny as neonate or adult with bone marrow from Des.TCR transgenic mice. Eight weeks after transfer the lymphoid compartments of both recipients were found to be efficiently repopulated with donor T cells. Induction of tolerance in Des.TCR × 2.4Ker.Rag-$2^{-/-}$ chimeric mice was judged by acceptance of K^b-transfected, syngeneic P815 tumour cells. Tolerance could only be observed in the neonatally reconstituted chimeric mice and was not detectable in the adult recipients. Therefore we have to conclude that, in contrast to the neonate, the adult environment of (Des.TCR × 2.4Ker-K^b)F1 mice does not allow tolerance induction to antigens expressed on keratinocytes in T cells newly exported from the thymus. This could be due to differential antigen expression in the neonatal compared to the adult animal. However, it should be noted that K^b is strongly expressed on hair follicles in the adult skin allowing 2.4Ker-K^b skin graft rejection by syngeneic K^b-negative mice. Alternatively, these results could be based on different migration behaviour of naïve T cells in neonatal and adult animals. Indeed, it has been recently reported that naïve T cells migrate into tissues of sheep embryos whereas such a migration was not observed in the adult sheep (Kimpton et al 1995).

Conclusions

The studies described here show that peripheral tolerance induction can be based on highly dynamic processes. Deletion of anti-K^b T cells either in the thymus (by undetectable amounts of antigen) or in the periphery can certainly not be the only mechanism to explain the observed tolerance to K^b in (Des.TCR × 2.4Ker-K^b)F1 mice. The lack of tolerance in the Des.TCR × 2.4Ker-K^b.Rag-2^- adult chimeras postulates regulatory events being responsible for the maintenance of tolerance in the adult (Des.TCR × 2.4Ker-K^b)F1 animals. This is in agreement with the observed heterogeneity in the activation marker expression and lymphokine secretion among the $CD8^+$ Des.TCR^+ T cells of the tolerant double-transgenic mice. The anti-K^b T cells secreting IL-2 and IFN-γ upon antigen stimulation may represent the naïve T cells which have matured in the adult animal and therefore could not encounter the K^b antigen. The IL-10- and IL-4-secreting $CD8^+$ Des.TCR^+ T cells might be the ones which have seen the K^b antigen in the neonatal phase, need the presence of antigen to maintain their self-renewal and to control the reactivity of the naïve anti-K^b T cells. The nature of this regulatory process remains to be identified.

Acknowledgements

We thank Mrs. Copson for the preparation of the manuscript. This work was in part supported by the Biotechnology Programme BIO4-CT96-0077, G.J.H./B.A. and DFG Ar 152/3-1, B.A.

References

Alferink J, Schittek B, Schönrich G, Hämmerling GJ, Arnold B 1995 Long life span of tolerant T cells and the role of antigen in maintenance of peripheral tolerance. Int Immunol 7:331–336

Arnold B, Schönrich G, Hämmerling GJ 1993 Multiple levels of peripheral tolerance. Immunol Today 14:12–14

Bertolino P, Heath WR, Hardy CL, Morahan G, Miller JF 1995 Peripheral deletion of autoreactive $CD8^+$ T cells in transgenic mice expressing H-$2K^b$ in the liver. Eur J Immunol 25:1932–1942

Ferber I, Schönrich G, Schenkel J, Mellor A, Hämmerling GJ, Arnold B 1994 Levels of peripheral T cell tolerance induced by different doses of tolerogen. Science 263:674–676

Groux H, Bigler M, de Vries JE, Roncarolo MG 1996 Interleukin-10 induces a long-term antigen-specific anergic state in human $CD4^+$ T cells. J Exp Med 184:19–29
Kimpton WG, Washington EA, Cahill RNP 1995 Virgin $\alpha\beta$ and $\gamma\delta$ T cells recirculate extensively through peripheral tissues and skin during normal development of the fetal immune system. Int Immunol 7:1567–1577
Limmer A, Sacher T, Alferink J, Nichterlein T, Arnold B, Hämmerling GJ 1998 A two-step model for the induction of organ-specific autoimmunity. In: Immunological tolerance. Wiley, Chichester (Novartis Found Symp 215) p 159–171
Mackay CR 1991 T cell memory: the connection between function, phenotype and migration pathways. Immunol Today 12:189–192
Miller JFAP, Morahan G 1992 Peripheral T cell tolerance. Annu Rev Immunol 10:51–69
Schönrich G, Kalinke U, Momburg F et al 1991 Downregulation of T cell receptors on self-reactive T cells as a novel mechanism for extrathymic tolerance induction. Cell 65:293–304
Schönrich G, Momburg F, Malissen M et al 1992 Distinct mechanisms of extrathymic T cell tolerance due to differential expression of self antigen. Int Immunol 4:581–590
Sprent J 1994 T and B memory cells. Cell 76:315–322
Tafuri A, Alferink J, Möller P, Hämmerling GJ, Arnold B 1995 T cell awareness of paternal alloantigens during pregnancy. Science 270:630–633
Zinkernagel RM, Pircher HP, Ohashi P et al 1991 T and B cell tolerance and responses to viral antigens in transgenic mice: implications for the pathogenesis of autoimmune versus immunopathological disease. Immunol Rev 122:133–171

DISCUSSION

Stockinger: I was glad to see you have used TCR transgenics on a Rag^- background. You stated that you couldn't see tolerance induction in the Rag mice, and I had the impression you were implying from that that we cannot work with Rag mice all the time because it is not the real situation: you need other cells if you believe in regulation. I would argue that if you modified your situation you would see it just as well in Rag mice. The reason you don't find it there is because you have an overwhelming frequency of K^b-specific cells coming in after the first wave of thymic emigrants in the neonate have become tolerized and turned into regulatory cells. These cells don't have a chance to influence all these late emigrants, whereas in the normal (Rag^+) situation the frequency of K^b-specific cells is low enough that the regulatory influence is visible. If you let your Rag mice go through the first wave of thymic emigration, assuming that in the neonatal period they have access to the K^b on the tip of the tongue, you then take the thymus out, look at that population (I would imagine it's one that might secrete IL-10 with a partial tolerance phenotype), then you could inject in small numbers of competent mature cells from Rag^- transgenic mice and see if you tolerized those. It doesn't make sense that you invoke a regulatory population unrelated to your K^b-specific transgenic T cells but nevertheless still recognizing K^b. The specificity is there, but you propose it comes from a population that has nothing to do with your transgenic T cell population.

Mitchison: Let's divide that comment into two bits. First your hypothesis, and then the experiment that you suggest to test it. As far as the hypothesis is concerned, is that really all that different from what Bernd Arnold was saying?

Arnold: I'm missing one piece of information: are you talking about the normal double-transgenic Rag mice?

Stockinger: The double transgenics.

Arnold: The double-transgenic Rag mice are able to reject the K^b $B7^+$ tumour. In these mice I only have $CD8^+$ clonotype$^+$ cells. I do not follow you when you say we have different waves: this is normal development in these mice.

Mitchison: Brigitta Stockinger's suggestion is that this is because that the Rag mice don't have the normal cells in them which are needed to mediate an interactive tolerance.

Stockinger: No, that's not what I'm saying. Bernd Arnold says that they don't have normal cells. I'm saying they have proportionally too few regulatory cells within their own transgenic population to cope with the mass of monospecific cells that are coming out of the thymus as the adult wave that isn't exposed to the antigen. It is a simple frequency problem. In a normal situation you have too few of these clonotype-specific cells, so they can be suppressed easily. In the Rag situation, all of them are of that type and they are not suppressed.

Arnold: There are not that few. In the mice with the normal rest repertoire, there is a high percentage of $CD8^+$ clonotype$^+$ cells.

Stockinger: There are not as many as in the Rag mouse, because you have other cells diluting them.

Arnold: Yes, but if you look at a normal repertoire we have about 20% $CD8^+$ clonotype$^+$ cells.

Stockinger: I'm saying that 20% is suppressible, but 80% is not.

Mitchison: I don't think that you've convinced us, Brigitta, that there's much difference between yours and Bernd's hypotheses.

Stockinger: The normal cells are the difference. I don't believe in normal cells having anything to do with that.

Abbas: Because of the K^b specificity. The argument is that the normal cells ought not to be K^b-specific, so why are they involved in this regulation process?

Arnold: You are talking about the rest repertoire in the Des.TCR mice. Do you doubt that this rest repertoire can influence an anti-K^b response?

Abbas: Precisely, because they shouldn't be K^b specific.

Arnold: Of course there are K^b-specific cells in the rest repertoire, because they have normal third-party reactivity. These mice reject $H\text{-}2^s$ skin grafts.

Stockinger: This repertoire is greatly reduced in a transgenic mouse.

Arnold: It is not neglectable.

Jenkins: There is a precedent in Doug Hanahan's work, where he produced mice that contained a transgenic TCR that was autoreactive in a mouse where autoantigen was expressed (Förster et al 1995). When the transgenic T cells accounted for 95% of the T cells he saw no tolerance, and when they accounted for 10% he saw good peripheral tolerance. So purely on the basis of the number of reactive cells you can argue that the peripheral tolerance mechanism can be overwhelmed.

Abbas: The experiment you proposed, which involves gradually reconstituting the Rag mice with normal cells and seeing what happens, is a good one. But which normal cells do you mean?

Stockinger: I was suggesting reconstituting thymectomized Rag^- transgenic mice with small numbers of T cells from the same type of Rag^- transgenic donor to see whether one and the same transgenic population itself can turn into regulatory cells that can tolerize other members of the population. I don't believe that an unrelated 'normal' population needs to be invoked for that effect, but the way the experiment has been done so far could now show this potential for regulation within one and the same T cell population.

Arnold: But this is just the question about whether the Rag double-transgenic mice are tolerant or not. However, this is not the major point I wanted to make. How can we resolve the discrepancy that in the adult T cells do not go into tissues, and we say we only have tissue-specific expression and we do have tolerance? The message I tried to give you is that apparently there is a solution. In the adult, naïve T cells don't go into the tissues. Accordingly, in the adult chimeras we don't have tolerance. However, in the neonate or in the embryo, naïve cells can go into tissues (Kimpton et al 1995). If we have expression in the neonatal phase, indeed we do have tolerance.

Healy: Your finding that transgenic T cells enter neonatal but not adult skin is very similar to models for autoimmune bullous skin diseases. Passive transfer of basement membrane-specific antibodies causes robust immunofluorescent staining and epidermal blisters in neonatal but not adult skin, despite presence of the antigen in both skin types. Access of the immune system to cutaneous antigens appears to be a general phenomenon that is developmentally regulated.

Jenkins: You said that the double-transgenic mice are tolerant but tolerance can be reversed in the presence of IL-2, and yet you have shown that the mice can contain cells making IL-2. How does this all fit together?

Arnold: We have a balance between these cells, which if you stimulate with an antigen will make IL-2 and IFN-γ, and other cells which react differently. Under normal circumstances, if you come with a normal stimulus, the balance is towards the other cells. Therefore you do have tolerance. However, if you give IL-2 plus antigen, of course you will expand the cells capable of producing IL-2, and therefore you tip the balance and can easily break tolerance.

Jenkins: So it's a quantitative matter to do with the amount of IL-2 that's there?

Arnold: I would assume so. You need the IL-2 plus the antigen. As Günter Hämmerling showed, if you take the P815 cells with only IL-2 you don't see the effect.

Jenkins: Am I right in thinking that you are measuring the IL-2 15 days after giving the tumour?

Arnold: We can also measure it earlier. I showed an extreme case because there we have so many cells: this was just to make the point that apparently we have expansion with time in the presence of the antigen.

Jenkins: When we've looked at that, in the case where we can give the antigen at a very specific time and look at the production of IL-2 by antigen-specific T cells *in vivo*, it's very transient—it is made over a period of one day. If this is also true in this system, then the fact that you've got a lot of cells making IL-2 suggests to me that the cells you detect at any given time are cells very recently stimulated by the antigen. So are you looking at cells that are actually in the process of being tolerized via some kind of aborted activation, or do you think these T cells are actually responding chronically to self antigen?

Arnold: We have no kinetics on a single-cell population, so we cannot say that a particular cell which currently makes IL-2 and IFN-γ, won't make IL-2 and IFN-γ in another 10 h. Somehow we have to accommodate the fact that these mice reject the tumour. To me, it seems likely that the cells which react in production of IL-2 and IFN-γ are the ones which reject the tumour.

Mitchison: Surely rejecting a tumour is an enormous antigenic stimulus compared with that of a small dose of ovalbumin.

Is this a new development in your thinking? I've been following the talks that you have given over the last few years quite closely, and it seems to me that this is a whole new area that you're getting into of regulatory T cells and interactions with cytokines, whereas before you were thinking of things very much in terms of an induced state of non-reactivity in these chimeras.

Arnold: You are correct. At the beginning we saw that the mice are tolerant and cells are present. By analogy to the *in vitro* studies, we thought they must be somehow anergic. But now we see there is heterogeneity among the cells in regard to activation markers and lymphokines, and we see that the adult chimeras are not tolerant. On the basis of these results we have to postulate regulatory events responsible for the maintenance of peripheral tolerance in the adult double-transgenic mice.

References

Förster I, Hirose R, Arbeit JM, Clausen BE, Hanahan D 1995 Limited capacity for tolerization of $CD4^+$ T cells specific for a pancreatic β cell neo-antigen. Immunity 2:573–585

Kimpton WG, Washington EA, Cahill RNP 1995 Virgin $\alpha\beta$ and $\gamma\delta$ T cells recirculate extensively through peripheral tissues and skin during normal development of the fetal immune system. Int Immunol 7:1567–1577

T lymphocyte-mediated control of autoimmunity

Ethan M. Shevach, Angela Thornton and Elisabeth Suri-Payer

Laboratory of Immunology, National Institute of Allergy and Infectious Diseases, National Institutes of Health, Bethesda, MD 20892, USA

Abstract. Autoreactive T cells can be readily identified in the peripheral lymphocyte pool of both humans and experimental animals. Peripheral tolerance may be maintained by regulatory/suppressor T cells which prevent the activation of autoantigen-specific cells. Mice thymectomized on day 3 of life (d3Tx) develop a wide spectrum of organ-specific autoimmune diseases. Reconstitution of d3Tx mice with $CD4^+$ $CD25^+$ T cells from normal mice prevents the development of disease. Similarly, $CD4^+$ $CD25^+$ T cells prevent the transfer of disease by autoantigen-specific cloned T cells derived from d3Tx mice. Thus, regulatory T cells can prevent both the induction and effector function of autoreactive T cells. *In vitro*, the $CD4^+$ $CD25^+$ population is anergic to stimulation through the T cell receptor (TCR) and suppresses the proliferative responses of normal $CD4^+$ $CD25^-$ cells by a contact-dependent mechanism. Suppression is not MHC-dependent, but requires activation of the $CD4^+$ $CD25^+$ population. The mechanism of suppression *in vivo* and the target antigen(s) of this unique regulatory population remain to be characterized.

1998 Immunological tolerance. Wiley, Chichester (Novartis Foundation Symposium 215) p 200–217

Studies utilizing transgenic mouse model systems have conclusively demonstrated that self-reactive T cells are frequently not deleted in the thymus and are not tolerized in peripheral lymphoid tissues, but rather persist in the periphery in a state of 'ignorance' or 'indifference' (Miller & Heath 1993). In general, it has been assumed that these cells are not responsible for the induction of autoimmune diseases because they do not come in contact with their target antigens which are expressed in non-lymphoid organs which are not normally 'visited' by the circulating lymphocyte pool. However, these cells with the latent propensity to induce autoimmune diseases have the potential to be activated when their target self antigens are released into the lymphoid system during the course of an infectious insult or when they are activated by cross-reactive epitopes present on infectious agents (Wucherpfennig & Strominger 1995). Since the incidence of autoimmune disease in the general population is low, it remains possible that other mechanisms have been developed to prevent the activation of the relatively large number of autoreactive cells present in normal animals and humans and thereby maintain a state of tolerance to self. A number of experimental models have been developed which suggest that regulatory T cell populations play a major role in preventing the development of organ-specific autoimmune diseases (Powrie et al 1994, Saoudi et al 1996). Studies over a number of years (Sakaguchi et al 1985) have suggested that a potent immunoregulatory T cell population is responsible for the prevention of autoimmune disease that is precipitated by thymectomy of mice on day 3 of life (d3Tx). In this report, we will characterize this immunoregulatory T cell population and demonstrate that it represents a unique lineage of professional 'suppressor' T lymphocytes.

Thymectomy of 3 day old mice is followed by the development of organ-specific autoimmune disease. The disease process is mediated by $CD4^+$ T cells and can be transferred by $CD4^+$ cells to *nu/nu* and SCID mice (Sakaguchi & Sakaguchi 1994). The syndrome is characterized by high titre antibodies to the involved organs and by an inflammatory infiltrate. In an affected mouse, one or more organs may be involved and these most frequently include stomach, ovaries/testes, thyroid, pancreas and epididymis. It is likely that distinct antigens are involved in each organ as intrathymic injection of gastric parietal cells (Murakami et al 1993) or transgenic expression in the thymus (Alderuccio et al 1993) of a purported target antigen prevents the development of gastritis, but not oophoritis. These results suggest that the T cells that mediate this disease can be tolerized in the thymus and that the effector phase of the disease is mediated by multiple populations of cells with specificities for distinct target antigens in different organs.

One of the most interesting aspects of this disease is that $CD4^+$ T cells from euthymic animals can inhibit the development of disease when transferred into the d3Tx animals before 14 days of age (Sakaguchi & Sakaguchi 1994). Taken together, previous studies of this model have been interpreted as demonstrating that autoreactive T cells are selectively exported from the thymus early in life because the mechanisms responsible for the induction of tolerance intrathymically are not completely functional during the neonatal period. In fact, we have demonstrated that the thymus exports immature T cells that are enriched in the progenitors of auotreactive cells during the neonatal period (Bonomo et al 1994). However, in euthymic animals these T cells are normally controlled by a second distinct population of regulatory T cells that develops later in ontogeny or can recirculate back to the thymus and be deleted (Bonomo et al 1995). It has also been proposed that the regulatory T cells whose development is prevented by d3Tx are responsible for preventing autoimmunity throughout the normal life-span of the mouse. Thus, transfer of adult $CD4^+$ T cells that express low levels of the CD5 antigen to *nu/nu* recipients results in a spectrum of disease in susceptible strains which closely resembles that seen post-d3Tx. More importantly, co-transfer of the $CD5^{hi}$ cells prevents the development of disease (Sakaguchi et al 1985). More recent studies by Sakaguchi and associates (Sakaguchi et al 1995, Asano et al 1996) have demonstrated that the immunoregulatory T cell population can be more precisely identified as those $CD4^+$ T cells which co-express CD25 (the interleukin 2 receptor α chain) which is approximately 10% of the peripheral $CD4^+$ T cell pool and considerably smaller that the $CD4CD5^{hi}$ population which comprises about 80% of the $CD4^+$ T cell pool.

Although these studies strongly suggest that regulatory T cells are involved in the control of potentially autoreactive T cells, one complicating feature of all of these studies is that autoimmunity is only observed in lymphopenic environments (post-d3Tx or post-transfer to immunodeficient mice). Indeed, we have proposed (Bonomo et al 1995) that the 'empty space' created by an 80% decrease in the number of peripheral T cells in the d3Tx animal will enhance the frequency of interactions between autoreactive T cells and antigen-presenting cells (APCs), and thereby lead to the priming of autoreactive effector populations. The altered homeostasis created by the thymectomy would be specific for different subpopulations of T cells as only $CD4^+$ T cells can inhibit the induction of disease post-d3Tx.

Characterization of the $CD4^+$ T effector cells in post-d3Tx autoimmunity

Resolution of these complex questions and distinctions between these models requires the use of both homogeneous populations of effector T cells which recognize the autoantigens and the development of experimental systems which will allow one to separate the function of T suppressor cells from the disordered status of the immune system in the lymphopenic state. The serum of d3Tx BALB/c mice contains high titre antibodies (Gleeson & Toh 1991) to the gastric parietal cell proton pump, the H/K ATPase α and β chains, and a number of studies had indirectly suggested that the H/K ATPase was also the target of $CD4^+$ T cells in this disease. In order to characterize the effectors which mediate autoimmune gastritis, we prepared H/K ATPase-enriched preparations of parietal cell microsomes and also purified the enzyme from these preparations by lectin affinity chromatography (Suri-Payer et al 1996). Our initial attempts to demonstrate reactivity of peripheral lymph node cells from d3Tx animals to the antigen were unsuccessful; however, when we tested lymph node cells in the immediate proximity of the stomach, vigorous MHC class II-restricted T cell proliferative responses were observed (Table 1). Apparently the precursor frequency of the autoantigen-specific T cells was too low in the peripheral nodes to allow detection of a response. There was an excellent correlation between the presence of H/K ATPase-reactive T cells, the titre of anti-parietal cell antibody

TABLE 1 Responses of lymph node populations to H/K ATPase

Lymph node	*Stimulation index*
d3Tx—gastric	12
d3Tx—pancreatic	7
d3Tx—mesenteric	3.5
d3Tx—peripheral	2
Control—gastric	1.5

Lymph node cells from d3Tx animals or from controls were cultured with enriched preparations of H/K ATPase (Suri-Payer et al 1996). Results are expressed as the stimulation index (counts per minute (CPM) with antigen/CPM in the absence of antigen).

(PCAb), and pathological evidence of gastritis. H/K ATPase-reactive T cells could first be detected as early as five weeks after thymectomy.

We have recently established several cloned lines of H/K ATPase-specific T cells by stimulating gastric lymph node cells from anti-PCAb-positive mice with rabbit microsomes and then cloned the cell lines by limiting dilution. The fine specificity of the clones was then tested with insect cell membranes prepared from baculovirus expressing either the H/K ATPase α and β chains together or individually. One clone, TxA-23, has been analysed in depth (Table 2) and clearly responds specifically to determinants expressed on the α chain. We have yet to detect a clone which responds to the β chain, even though the determinant responsible for the initiation of disease may be expressed on that chain (Alderuccio et al 1993). Clone TxA-23 is a typical Th1 clone which produces interferon-γ, but not interleukin (IL)-4 or IL-10, following antigen stimulation *invitro* (data not shown). The availability of the cloned population of effector cells allowed us to test whether the clone could transfer disease *in vivo*. Graded numbers of TxA-23 cells were transferred to normal BALB/c or C.B-17 SCID mice, or *nu*/*nu* mice on a BALB/c background. While the two immunodeficient mouse strains were readily susceptible to induction of disease by low numbers of transferred cloned T cells, normal BALB/c mice were quite resistant (Table 3). It should be noted that clone TxA-23 was incapable of helper activity for antibody production so no anti-PCAb was detected in the adoptive recipient and disease is measured solely as extent of pathological destruction of the stomach.

CD4^{+} CD25^{+} T cells are members of a unique lineage of suppressor cells

The ability of the cloned T cells to induce disease in *nu*/*nu* and SCID, but not normal, mice is consistent with either the presence of an 'empty space' in the former or the presence of immunoregulatory T cells in the latter. To distinguish

TABLE 2 Reactivity of clone TxA-23 to H/K ATPase

Stimulus	*$[^3H]$-TdR incorporation*
Media	1320
Rabbit H/K ATPase	23847
Baculovirus α chain	8849
Baculovirus β chain	639
Baculovirus $\alpha+\beta$ chains	8593

TxA-23 cells were stimulated with rabbit microsomes enriched for H/K ATPase or with membranes from insect cells transfected with rat H/K ATPase α, β, or $\alpha+\beta$ chains. $[^3H]$-TdR incorporation was determined after 96 h culture.

TABLE 3 Susceptibility of different mice to the induction of gastritis by TxA-23 cells

Transfer	*Recipient*	*% Gastritis*
TxA-23	BALB/c	15
TxA-23	BALB/c *nu/nu*	70
TxA-23	C.B-17 SCID	77
TxA-23	TCR Transgenic SCID	100
TxA-23+normal spleen	BALB/c *nu/nu*	0

TxA-23 cells were transferred to 8-week-old recipients or coinjected with normal BALB/c spleen cells containing $CD4^+$ $CD25^+$ cells. Gastritis was evaluated histologically after 6 weeks.

between these possibilities, we transferred the TxA-23 cells to SCID mice which expressed a transgenic TCR specific for a peptide derived from ovalbumin. Although these mice had a relatively normal number of $CD4^+$ T cells in their lymphoid tissues, they were very susceptible to disease induced by TxA-23 cells (Table 3). Thus, filling up of the lymphoid tissues of a mouse with a monospecific population of T cells was insufficient to prevent disease and strongly suggests that it is the regulatory T cells in normal animals that prevent disease. Indeed, the number of $CD4^+$ $CD25^+$ TCR transgenic T cells in the SCID mice was reduced by 70% compared with those present in normal mice. To test this hypothesis directly, we co-transferred TxA-23 cells and normal spleen cells containing $CD4^+$ $CD25^+$ T cells to *nu/nu* mice. Marked suppression of the induction was seen in the co-transfer study (Table 3).

The results of these studies are consistent with the possibility that the resistance of normal mice to the induction of autoimmune disease by the H/K ATPase-specific T cell clone is secondary to the presence of the immunoregulatory $CD4^+$

$CD25^+$ population. However, these studies do not address the issue of whether the $CD4^+$ $CD25^+$ T cells represent the normal population of activated T cells which are responding to environmental antigens or whether these cells are members of a unique lineage of suppressor cells. Reconstitution of d3Tx mice with normal spleen cells completely prevented the subsequent development of autoimmune gastritis (Table 4), while reconstitution with spleen cells which had been depleted of $CD4^+$ $CD25^+$ T cells had no therapeutic effect. As expected from the studies described above, cells from the TCR transgenic SCID mouse also were unable to prevent the induction of disease; more importantly, no amelioration of disease was seen when the d3Tx mice which had been reconstituted with the TCR transgenic cells were immunized with ovalbumin (Table 4). As immunization with antigen induced CD25 expression on ~50% of the TCR transgenic T cells, these results strongly support the view that the $CD4^+$ $CD25^+$ T cells present in normal animals, but absent in the TCR transgenic SCID, are members of a unique lineage of cells.

Characterization of the immunoregulatory functions of the $CD4^+$ $CD25^+$ T cells *in vitro*

We have developed reproducible methods for the purification of the $CD4^+$ $CD25^+$ T cells from normal animals (Fig. 1) and have evaluated the responsiveness of the purified cells to a variety of stimuli *in vitro*. If this population of cells represented T cells which had been activated *in vivo*, one might have predicted that they might be hyper-responsive to stimulation *in vitro*. However, the $CD4^+$ $CD25^+$ T cells were completely non-responsive to stimulation with soluble anti-CD3 and plate-bound anti-CD3, and also failed to respond to high concentrations of IL-2; the population could proliferate normally when challenged with the combination of phorbol ester and calcium ionophore which is independent of the TCR (Fig. 2). The lack of

TABLE 4 $CD4^+$ $CD25^+$ TCR transgenic T cells do not inhibit d3Tx-induced gastritis

Reconstitution	*% Gastritis*
None	54
Normal spleen	0
$CD4^+$ $CD25^-$	58
TCR Transgenic	47
TCR Transgenic/immunize	59

d3Tx BALB/c mice were left untreated or were reconstituted with normal spleen cells, CD25-depleted normal spleen cells, or spleen cells from TCR transgenic SCID mice. Some of the recipients of the transgenic T cells were immunized with ovalbumin on the day of cell transfer. Mice were examined for histological evidence of gastritis at 8 weeks of age.

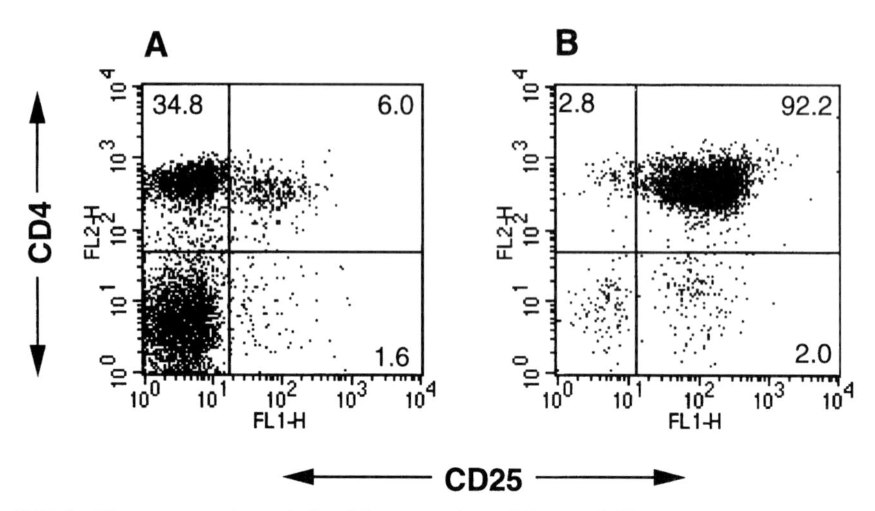

FIG. 1 Flow cytometric analysis of the expression of CD4 and CD25 on unseparated lymph node cells (Panel A) and on purified $CD4^+$ $CD25^+$ cells (Panel B). Cells were stained with PE–anti-CD4 and FITC–anti-CD25 and analysed on a FACScan (Becton Dickinson). We purified the $CD4^+$ $CD25^+$ population from normal lymph node cells by enriching for $CD4^+$ T cells using T cell purification columns followed by $CD8^+$ T cell depletion using anti-CD8 magnetic beads. $CD4^+$ $CD25^+$ cells were then purified by incubation with biotin anti-CD25 and strepavidin-FITC followed by positive selection with anti-FITC magnetic microbeads.

response of the $CD4^+$ $CD25^+$ population prompted us to examine whether they would suppress the response of $CD4^+$ $CD25^-$ T cells to stimulation. Marked suppression of the response of the $CD4^+$ $CD25^-$ to stimulation with soluble anti-CD3 in the presence of accessory cells was observed, but no inhibition of the response to plate-bound anti-CD3 was seen (Table 5); suppression was not restricted by the MHC as $CD4^+$ $CD25^+$ T cells from C57BL/6 mice suppressed the response of $CD4^+$ $CD25^-$ T cells from BALB/c mice. Most importantly, suppression required activation of the $CD4^+$ $CD25^+$ population as the $CD25^+$ cells were incapable of suppressing the responses of TCR transgenic mice to ovalbumin, but could readily suppress their response to anti-CD3 (Table 5). We have also demonstrated that the suppression observed in the co-cultures was not mediated by the secretion of suppressive cytokines (IL-4, IL-10 or TGF-β) by the $CD4^+$ $CD25^+$ T cells and required cell contact between the suppressor population and the responders (data not shown).

What is the physiological ligand of the $CD4^+$ $CD25^+$ population?

Taken together, the results of these *invivo* and *invitro* studies support the hypothesis that the $CD4^+$ $CD25^+$ population represents a unique lineage of suppressor T cells.

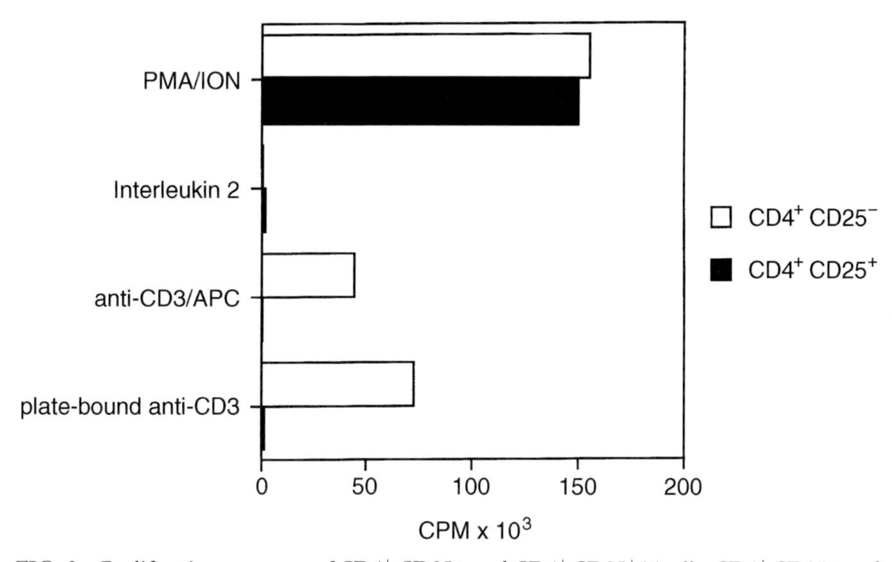

FIG. 2 Proliferative responses of $CD4^+$ $CD25^-$ and $CD4^+$ $CD25^+$ T cells. $CD4^+$ $CD25^-$ and $CD4^+$ $CD25^+$ populations were purified from lymph node as described in the legend to Fig. 1 and cultured (5×10^4/well) in the presence of the indicated stimuli and T-depleted spleen cells (5×10^4/well) as accessory cells. PMA/ION, phorbol myristate acetate/calcium ionophore; CPM, counts per minute.

Although our *in vitro* studies have thus far been performed with a polyclonal activator, anti-CD3, while our *in vivo* experiments have been executed with an autoantigen-specific T cell clone, we would like to combine the data from these two model systems in a speculative discussion as to the possible ligand which is recognized by the suppressor population. This model must also incorporate our preliminary *in vitro* observations that the suppressor population must be activated to exert its biological functions and mediate suppression by a cell contact-dependent mechanism.

Previous studies on the function of immunoregulatory T cells in the d3Tx model have focused on the possibility that the suppressor population recognizes the same target antigen as the autoimmune effector cells. The suppressor cell could then compete for a limited supply of antigen and block the effectors or perhaps under certain conditions secrete suppressor cytokines. Such a model would easily explain most of our experimental observations and one could easily argue that the ability of the $CD4^+$ $CD25^+$ T cells to suppress *in vitro* is secondary to polyclonal activation of both the suppressors and effectors with the suppressors dominating. The major problem with this model is that it would require a large, distinct suppressor population that would recognize the same universe of peptide–MHC class II complexes as the effector cells. However, it is also possible that the $CD4^+$ $CD25^+$

TABLE 5 Inhibition of the responses of CD4⁺ CD25⁻ T cells by CD4⁺ CD25⁺ T cells

Inhibitor	*Stimulus*	*[^{3}H]-TdR incorporation*
(A) BALB/c CD4$^+$ CD25$^-$		
—	Anti-CD3/APC	75196
CD4$^+$ CD25$^+$	Anti-CD3/APC	9495
CD4$^+$ CD25$^+$ (B6)	Anti-CD3/APC	17057
—	Plate-bound anti-CD3	78022
CD4$^+$ CD25$^+$	Plate-bound anti-CD3	83383
(B) TCR transgenic CD4$^+$ CD25$^+$		
—	OVA	40256
CD4$^+$ CD25$^+$	OVA	47995
—	Anti-CD3/APC	51799
CD4$^+$ CD25$^+$	Anti-CD3/APC	3909

CD4$^+$ CD25$^-$ T cells (5×10^4) from normal BALB/c, TCR transgenic mice, or C57BL/6 mice (line 3; the response of the C57BL/6 CD4$^+$ CD25$^+$-cells alone was ~75 000) were cultured alone or in the presence of CD4$^+$ CD25$^+$ T cells (12.5×10^4) from BALB/c mice. [^{3}H]-TdR incorporation was determined after 72 h culture. OVA, ovalbumin.

population contains T cells with a restricted repertoire of receptors which recognize dominate 'suppressor epitopes' within a complex protein antigen (Adorini et al 1979). Activation of the suppressor population by these peptides could easily lead to bystander suppression of CD4$^+$ effector cells recognizing other epitopes on the same antigen at the same local site.

An alternative to an antigen-specific suppressor population would be an anti-idiotype suppressor cell (Kumar & Sercarz 1993). It would be easier to understand the function of such a population if they were exclusively CD8$^+$ T cells as they could then recognize peptides derived from the TCR of the effector cells expressed in association with MHC class I molecules on the effector cells themselves. However, the suppressor population in this model is exclusively CD4$^+$ and the effector cells do not express MHC class II antigens; thus, the suppressor cell would need to recognize shed peptide antigens derived from the TCR of the effector cells on the surface of an APC. How such cells would then mediate their suppressive functions is not clear, but is possible that they might kill the physically adjacent effector cells by a Fas/Fas-ligand dependent mechanism.

Lastly, we would like to propose a distinct model for the function of the CD4$^+$ CD25$^+$ suppressor lineage. A major tenet of this hypothesis is that the regulatory T cell population is autoreactive and is specific for ubiquitously expressed autoantigens. Although one might immediately argue that the existence of such a population of autoreactive T cells in the peripheral lymphoid tissues is

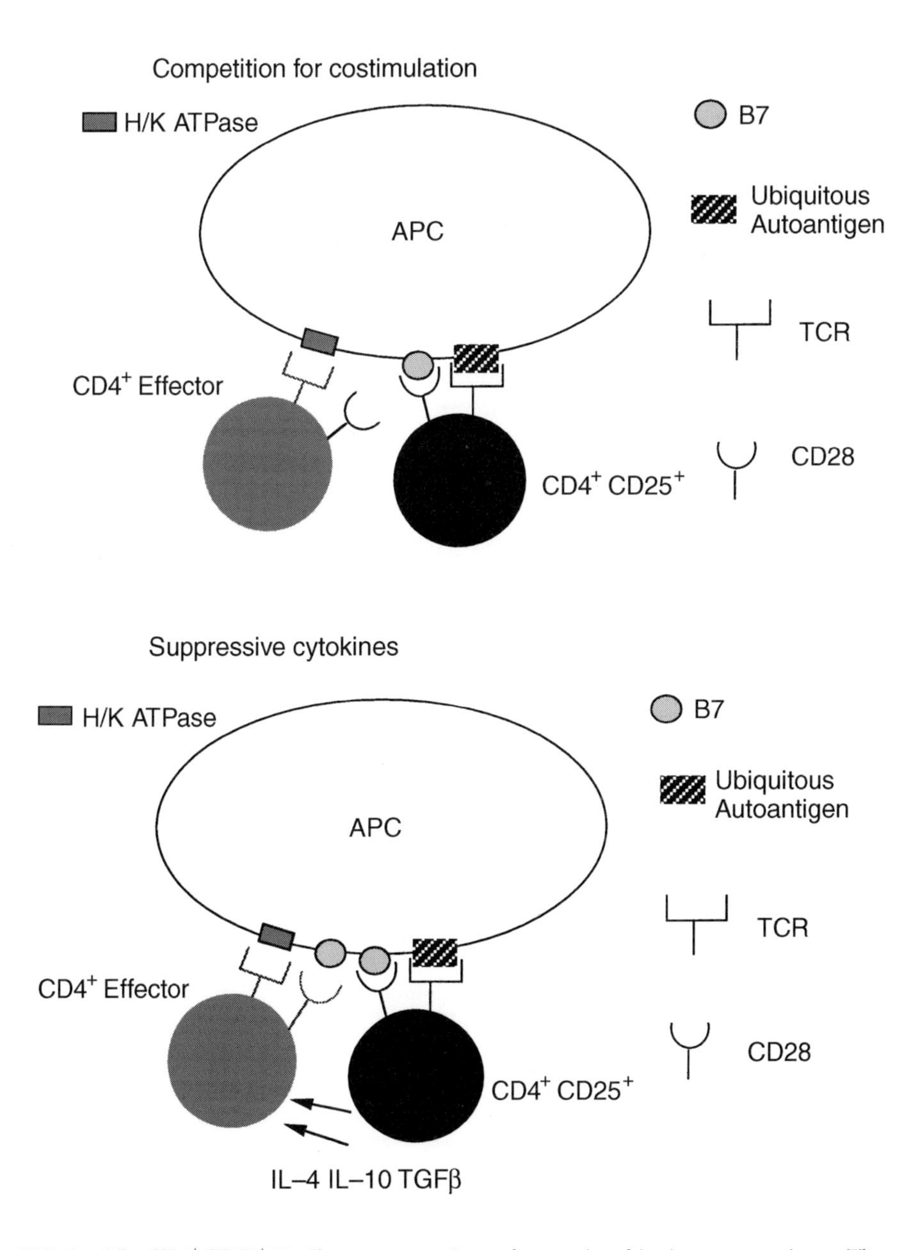

FIG. 3 The CD4+ CD25+ T cells are autoreactive and recognize ubiquitous autoantigens. The CD4+ CD25+ cells inhibit the activation of the CD4+ effector T cells by competing for co-stimulatory signals (*top panel*). The CD4+ CD25+ T cells inhibit activation by secreting suppressor cytokines (*bottom panel*).

incompatible with normal immunological function (or life), a second tenet of this model is that such autoreactive T cells must have down-regulated their capacity to differentiate into Th1 effector cells. Although the process of negative selection in the thymus is regarded as being highly efficient, it remains possible that some T cells recognize 'self-antigens' in a slightly altered fashion. Rather than undergoing apoptosis and elimination in the thymus, such cells would alter their TCR signalling properties and thus develop into suppressor cells. One obvious mechanism might involve a fixed differentiation along the Th2 pathway of suppressor cytokine production; alternatively, such cells could alter their capacity to respond to co-stimulatory molecules such that the CD28 molecule on their cell surface binds B7, but fails to transmit a signal which allows clonal expansion (Fig. 3). In the latter case, the suppressor population might inhibit the activation of Th1 autoantigen-specific effector cells which are recognizing their target autoantigen in the immediate microenvironment by preventing them from receiving the necessary co-stimulatory signals. In order to maintain normal immunological functions, such suppressor populations would only be capable of inhibiting T cells expressing receptors of relatively low affinity for their target antigens (autoantigens) and would have no effect on T cells which express receptors of relatively high affinity which recognize foreign antigens (pathogens).

Summary

The notion of 'suppressor T cells' has had a rather a jaded history in the immunological literature over the past 25 years. We caution that all the models proposed above must obey the now well-established rules that govern TCR recognition and function. Irrespective of the nature of the physiological ligand recognized by the $CD4^+$ $CD25^+$ population, our studies strongly support the view that they represent a unique functional population which should be subject to modulation *in vivo*. The identification of agents that would enhance their suppressive functions might prove to be of therapeutic benefit in many autoimmune diseases. Although complete elimination of the $CD4^+$ $CD25^+$ T cells might lead to the development of autoimmunity (Taguchi & Takahashi 1996), a more controlled decrease in their number/function might be used to enhance immune responses to tumour-associated antigens or facilitate immune responses to vaccination.

References

Adorini L, Harvey MA, Miller A, Sercarz EE 1979 Fine specificity of regulatory cells. II. suppressor and helper T cells are induced by different regions of hen egg white lysozyme in a genetically nonresponder mouse strain. J Exp Med 150:293–306

Alderuccio F, Toh BH, Tan SS, Gleeson PA, van Driel IR 1993 An autoimmune disease with multiple molecular targets abrogated by the transgenic expression of a single autoantigen in the thymus. J Exp Med 178:419–426

Asano M, Toda M, Sakaguchi N, Sakaguchi S 1996 Autoimmune disease as a consequence of developmental abnormality of a T cell subpopulation. J Exp Med 184:387–396

Bonomo A, Kehn PJ, Shevach EM 1994 Premature escape of double-positive thymocytes to the periphery of young mice. Possible role in autoimmunity. J Immunol 152:1509–1514

Bonomo A, Kehn PJ, Shevach EM 1995 Post-thymectomy autoimmunity: abnormal T cell homeostasis. Immunol Today 16:61–66

Gleeson PA, Toh BH 1991 Molecular targets in pernicious anaemia. Immunol Today 12: 233–238

Kumar V, Sercarz E 1993 T cell regulatory circuitry: antigen-specific and TCR-idiopeptide-specific T cell interactions in EAE. Int Rev Immunol 9:287–297

Miller JFAP, Heath WR 1993 Self-ignorance in the peripheral T-cell pool. Immunol Rev 133:131–150

Murakami K, Maruyama H, Nishio A et al 1993 Effects of intrathymic injection of organ-specific autoantigens, parietal cells, at the neonatal stage on autoreactive effector and suppressor T cell precursors. Eur J Immunol 23:809–814

Powrie F, Leach MW, Mauze S, Menon S, Caddle LB, Coffman RL 1994 Inhibition of Th1 responses prevents inflammatory bowel disease in *scid* mice reconstituted with $CD45RB^{hi}CD4^{+}$ T cells. Immunity 1:553–562

Sakaguchi S, Fukuma K, Kuribayashi K, Masuda T 1985 Organ-specific autoimmune disease induced in mice by elimination of T cell subset. I. Evidence for the active participation of T cells in natural self-tolerance: deficit of a T cell subset as a possible cause of autoimmune disease. J Exp Med 161:72–87

Sakaguchi S, Sakaguchi N 1994 Thymus, T cell and autoimmunity: various causes but a common mechanism of autoimmune disease. In: Coutinho A, Kazatchkine M (eds) Autoimmunity: physiology and disease. Wiley-Liss, New York, p 203–227

Sakaguchi S, Sakaguchi N, Asano M, Itoh M, Toda M 1995 Immunologic self-tolerance maintained by activated T cells expressing IL-2 receptor α-chains (CD25). Breakdown of a single mechanism of self-tolerance causes various autoimmune diseases. J Immunol 155:1151–1164

Saoudi A, Seddon B, Heath V, Fowell D, Mason D 1996 The physiological role of regulatory T cells in the prevention of autoimmunity: the function of the thymus in the generation of the regulatory T cell subset. Immunol Rev 149:195–216

Suri-Payer E, Kehn PJ, Cheever AW, Shevach EM 1996 Pathogenesis of post-thymectomy autoimmune gastritis. Identification of anti-H/K adenosine triphosphatase-reactive T cells. J Immunol 157:1799–1805

Taguchi O, Takahashi T 1996 Administration of anti-interleukin-2 receptor antibody *in vivo* induces localized autoimmune disease. Eur J Immunol 26:1608–1612

Wucherpfennig KW, Strominger JL 1995 Molecular mimicry in T cell-mediated autoimmunity: viral peptides activate human T cell clones specific for myelin basic protein. Cell 80:695–705

DISCUSSION

Hafler: Have you looked at the T cell receptor (TCR) of the CD25 T cell population? Do these cells have an invariant TCR?

Shevach: When we stain the TCR on the CD25 population with a panel of anti-V_β antibodies, the distribution of V_β specificities is identical on both the $CD25^+$ and $CD25^-$ T cells.

Hafler: Can the cells suppress experimental autoimmune encephalomyelitis (EAE), for example by adoptive transfer?

Shevach: We haven't done those experiments yet, but this is one explanation for why myelin basic protein (MBP) TCR transgenics on a conventional background don't get EAE, but if you breed them to $Rag^{-/-}$ mice they rapidly and spontaneously develop disease.

Hafler: You said they are not MHC-restricted. Do you have any idea what else can trigger them beside anti-CD3?

Shevach: We haven't yet been able to separate the activation of the suppressor from the activation of the responder.

Mason: I would like to question your conclusion that in the 3 day old thymus the mouse puts out autoagressive cells but not a regulatory population. We've done some experiments transferring thymocytes from adult animals into our lymphopenic animals that otherwise get diabetes, looking at a very similar mechanism to you. We find that whether you get good suppression of the diabetes or not depends critically on the number of cells you transfer. We have analysed this in a mathematical model in which we take the view that we're transferring regulatory cells and cells that can cause disease at the same time. In the inoculum we have a modest excess of the regulatory population over the one that causes disease. The more cells you transfer the better your sampling becomes, so you ultimately end up with a situation where if you transfer enough cells you get almost complete protection, but if you transfer a much smaller number of cells you don't get so much protection because of the statistical variation between the regulatory population and the one that causes disease.

In the first 3 days of life, the mouse doesn't put out that many T cells, so your mouse looks like our rat that's been given a fairly small number of adult thymocytes: those do not protect against disease very well. Consequently, I'm not sure that you can argue from what you've seen that it is a question of what cells are actually put out by the thymus in the first 3 days of life.

Shevach: There are certain strains of mice that are totally resistant to this phenomenon. For instance, if you thymectomize C57BL/6 mice, they don't get any disease at all. I'm not arguing with you: I think what you say is quite reasonable. Perhaps the B6 mouse puts out these regulatory cells two days earlier than the BALB/c mouse. CD25 is a good marker in that sense.

Wraith: Ethan Shevach, you do not favour the model that inhibition is cytokine mediated because of your studies with antibodies that were supposed to neutralize cytokines *in vitro*. Are you absolutely confident that those antibodies really do neutralize all the cytokines?

Shevach: In other experiments they actually do. However, *in vitro* for 3 days is very different from *in vivo* for 6 weeks and it remains quite possible that cytokines are responsible for inhibition of disease *in vivo*. We are evaluating this possibility by reconstituting the d3Tx mice with $CD25^+$ cells from $IL\text{-}4^{-/-}$ and $IL\text{-}10^{-/-}$ mice. We are also attempting to inhibit the disease induced by the TxA-23 clone in nude mice with the same populations. Mice reconstituted with $CD25^+$ cells from normal donors will also be treated with anti-IL-4 and anti-IL-10 or perhaps anti-TGF-β in an attempt to abrogate suppression. In the experiments of Powrie and Coffman (Powrie et al 1996), in the inflammatory bowel disease model, mice are treated with anti-TGF-β for a long time and the read-out is after a long period. *In vitro* it is pretty clear: we've looked every possible way for the production of IL-4 and IL-10 (less so for TGF-β) and there is absolutely no evidence for that, but that is *in vitro*.

Wraith: Concerning the other model — the antigen-specific model — I don't see how you can use this to explain the 3 day thymectomy effect. Can I propose a further model to account for this, which is that there are regulatory cells specific for proteins which come up in inflammation, such as heat shock proteins. Could it be that those are the cells that are being generated in that period?

Shevach: So this is a sort of an anti-inflammatory cell of a general nature marked by this marker. That's also a possibility.

Allen: I would like to propose yet another model! This is simpler and less heretical. Like the CTLA-4 issue, what if the environment of the immune system is actually negative in nature, and because of positive selection you have an autoreactive storehouse of T cells? The general idea is that you're going to keep this negatively controlled. What you've revealed is that there are some cells that are doing this. It might not be a specific phenomenon, and all of these different systems we've heard basically revealing this phenomenon in some manipulated system. The idea is that it is not a neutral environment in which you are waiting for a positive signal, but the environment in general is a little negative because your T cells are always just a little autoreactive.

Shevach: What do they see? It's not every T cell: 90% of the T cells aren't regulatory cells, 10% are and we'll get that number down I suspect with better antibodies and markers. So it is a unique population.

Allen: The problem with the old suppressor cell is that you have to have one cell chasing another one around, which to me has problems conceptually. I'm just trying to propose a general mechanism for immune regulation.

Shevach: The easiest way to think about these cells is that they are specific for the same antigen as the autoreactive effector cells. In this model, the target antigen, the H/K ATPase, is never expressed intrathymically so T cells which recognize it are never subjected to negative selection. A large population of autoreactive effector is present, but in some manner a suppressor population with the same antigen specificity is generated which controls the effectors. Both populations could migrate to the gastric node and recognize the same peptide on the surface of dendritic cells, but the suppressor prevents the effector from being activated (perhaps by competition).

Stockinger: I quite liked Paul Gleeson's interpretation that the lymphopenic state you create somehow provides a stimulus: the mice get easily infected or something like that so that the presentation of self-antigen is enhanced to stimulate this autoreactive T cell population (Gleeson et al 1996). I wonder, can you put the whole effect down to a hierarchy in seeding of T cell areas in normal mice which is disturbed in lymphopenic mice? I thought you were nearly there when you put your DO11.10 OVA transgenic T cells into these mice. The mistake was that you activated them with complete Freund's adjuvant (CFA), which probably tipped the autoreactive cells so much over balance that they really exploded. Normally these might be competed out in an intact and healthy immune system.

Shevach: In the absence of adjuvant the DO11.10 cells did not suppress, presumably because they contain very few $CD25^+$ cells.

Stockinger: When you look at re-population and homing experiments, there are certain cells that replace others and autoreactive ones may be displaced very easily under normal circumstances. Perhaps you just perturbed this homeostasis?

Shevach: Let's put the role of infection in the d3Tx model to rest. Germ-free mice develop only a slightly lower incidence of gastritis than normal mice following thymectomy on day 3 of life (Murakami et al 1992). This should be contrasted with the other inflammatory bowel disease models ($CD45Rb^{high}$ to SCID, and IL-10- and IL-2-deficient mice) where infection in the bowel plays a prominent role. We thought that the DO11.10 to d3Tx experiment was a reasonable approach to study the effect of induction of $CD25^+$ cells *in vivo* following immunization. However, you are correct then when we immunized the animals with ovalbumin in CFA that some enhancement of disease was seen in the immunized animals.

Jenkins: Did you try OVA without adjuvant?

Shevach: No.

Lenardo: The simple prediction of your model would be that the IL-2 receptor (IL-2R)-positive cells are different at day 3 than in the adult, either in numbers or in some feature that you can measure. Have you compared them?

Shevach: We can detect $CD4^+$ $CD25^+$ T cells in the lymph nodes of animals as early as day 3 of life. This is in contrast to the findings of Sakaguchi's group who only examined spleen (Asano et al 1996). The percentage (10%) of $CD4^+$ cells which co-express CD25 is constant throughout life in euthymic animals. Paradoxically, post-d3Tx the percentage of $CD4^+$ $CD25^+$ T cells rises to 20% by day 10 of life and can reach as high as 30% of the $CD4^+$ population. I believe that in the 3dTx animal the $CD4^+$ $CD25^+$ population actually represent autoreactive effectors which could see a host of antigens besides the H/K ATPase. So far we have not been successful in isolating $CD4^+$ $CD25^+$ cells from 3 day old euthymic animals, although we would be interested in analysing their function.

Mitchison: Can I ask just you and Don Mason to clarify the issue between you. Don has been doing this type of experiment with different markers in the rat. Are the two systems the same, and could the protective cell in your T cell population, Don, which is a larger fraction, be these very cells which Ethan Shevach is looking at here?

Mason: There is not a numerical discrepancy. If we use all the markers in conjunction, we can get the total number of thoracic duct lymphocytes that actually mediate this effect down to about 3%.

Mitchison: Does your marker include the α chain of IL-2R?

Mason: We haven't done those experiments exhaustively, but we're looking in the thoracic duct lymph and not in the spleen: that could be different.

Mitchison: But is that marker available in the rat?

Mason: Yes.

Mitchison: It hasn't been done?

Mason: No, although we did a small experiment with that very early on, with only three animals. We didn't see any evidence that it was the $IL\text{-}2R^+$ cells. But it was a small experiment and it was not spleen, it was thoracic duct lymph. They may well change their phenotype.

What we would like to find is a marker that is unique to these cells. IL-2R is not. We have screened an awful lot of monoclonal antibodies looking for a unique marker.

Shevach: Do you actually believe these cells are antigen-specific?

Mason: We do not have conclusive data ourselves and I don't think anybody does, but there is a very good experiment that supports the idea that the T cells that prevent autoimmunity are specific for autoantigens. Peter McCullagh did an experiment a long time ago in which he ablated the thyroid of rats *in utero* by giving the pregnant female ^{131}I (see McCullagh 1996). When he transplanted syngeneic thyroid tissue into the offspring after they had reached adulthood, the grafts were rejected. The implication is that if the target tissue for an autoimmune response is not present, the animal is not tolerant to it. We cannot say at this stage that the deficiency in self-tolerance in this case is due to a lack of the regulatory cells that are present in normal animals and that can prevent autoimmune diabetes or thyroiditis in lymphopenic rodents. The hypothesis is a testable one, however, and we are in the process of carrying out the relevant experiments.

Shevach: A number of experiments have been done to address this issue, but the results are clearly not conclusive. It has been shown in the oophoritis model, that $CD4^+$ T cells from males are just as effective as $CD4^+$ T cells from females in preventing disease post-3dTx (Smith et al 1991). However, it is quite possible that the T cell repertoire of a male mouse would contain as many receptors specific for ovarian antigens as the T cell repertoire of a female mouse; so the issue is not resolved. A different approach has been used by Murakami et al (1993) who concluded that the suppressor cells were not antigen-specific because neonatal intrathymic injection of gastric parietal cells followed by thymectomy on day 5 of life prevented the development of gastritis, but suppressor cells capable of preventing disease could still be isolated from these animals.

Mitchison: Could I ask you a question on antigen specificity. Do you think there's any future in testing conjugates—for example, your peptide linked to ovalbumin—to see if you can get anything out of the $CD25^+$ cells from the ovalbumin T cell transgenics?

Shevach: I actually don't think the transgenic SCIDs have this lineage of cells at all. They have a few $CD25^+$ cells and these represent a different population. I wouldn't do that experiment.

Mitchison: Why do you exclude that possibility? The conjugate experiment works in other systems.

Shevach: I don't think that the DO11.10 SCID mouse contains T cells that can potentially recognize the linked antigen you propose we create.

Allison: This may be a dumb idea, but what's the possibility that these cells are just IL-2 sinks?

Shevach: We have obviously thought about this possibility. The best experiment to rule out IL-2 consumption is to demonstrate that the suppressor cells actually block IL-2 production at the mRNA level. Such studies are in progress and preliminary data suggests that this is the case. Curiously, induction of the IL-2 receptor is not blocked.

Kurts: Marc Jenkins' experiments showing T cell clustering suggest that activated $CD4^+$ T cells can kill the APC after a short time. So if we suppose that some of the $CD25^+$ $CD4^+$ T cell population are antigen-specific, could it be that they kill the APCs expressing peptides derived from the proton pump? Then pathogenic host T cells could not be activated to cause tissue damage.

Shevach: That's a perfectly good explanation. Again, they would go to the draining lymph node in the gut. There's very little antigen there; they wipe those few cells out. That is compatible with our data.

Lenardo: That's inconsistent with your observation that they are anergic: you said that they don't respond to TCR signals so they may not stick to the dendritic cell because they are not going to up-regulate the appropriate adhesion molecule.

Jenkins: We don't know that cluster assay really requires activation of the cell.

Lenardo: It should.

Shevach: In vivo this could take days to weeks. The *in vitro* experiments are nice but they're limited. The ability of the suppressor cells to kill the APC *in vivo* could take a long time to develop.

Lechler: There are two predictions that you could make if the civil servant model is valid. One is that there is going to be some IL-2 consumption. The other is that there's going to be competition for the APC surface. So how do these cells behave in terms of clustering?

Shevach: We haven't examined that yet. They aren't homogenous. We have to be able to grow them: this is going to be the key and it is not going to be easy.

Wraith: Are you sure that they don't respond in the presence of anti-CD28 to, say, pump protein?

Shevach: We haven't done that yet.

Mitchison: Are you making a transgenic for the TCR of your disease-inducing clone?

Shevach: That is already in progress. My prediction is that such a TCR transgenic on a conventional background will not develop disease, but that on a $Rag^{-/-}$ background severe disease will be seen.

References

Asano M, Toda M, Sakaguchi N, Sakaguchi S 1996 Autoimmune disease as a consequence of developmental abnormality of a T cell subpopulation. J Exp Med 184:387–396

Gleeson PA, Toh B-H, van Driel IR 1996 Organ specific autoimmunity induced by lymphopenia. Immunol Rev 149:97–125

McCullagh P 1996 The significance of immune suppression in normal self tolerance. Immunol Rev 149:127–153

Murakami K, Murayama H, Hosono M et al 1992 Germ-free condition and the susceptibility of BALB/c mice to post thymectomy autoimmune gastritis [letter]. Autoimmunity 12:69–70

Murakami K, Maruyama H, Nishio A et al 1993 Effects of intrathymic injection of organ-specific autoantigens, parietal cells, at the neonatal stage of autoreactive effector and suppressor T cell precursors. Eur J Immunol 23:809–814

Powrie F, Carlino J, Leach MW, Mauze S, Coffman RL 1996 A critical role for transforming growth factor β but not interleukin 4 in the suppression of T helper type 1-mediated colitis by CD45RB(low) CD4+ T cells. J Exp Med 183:2669–2674

Smith H, Sakamato Y, Kasai K, Kung TS 1991 Effector and regulatory cells in autoimmune oophoritis elicited by neonatal thymectomy. J Immunol 147:2928–2933

Final general discussion

Mitchison: My suggestion for this general discussion is that we should take the four main groups of molecules which have been the centre of attention during this meeting, and perhaps add to these the phosphatases which have been proposed here as being important in anergy, and address the following question about all of them: how can we fit these molecular systems into allograft acceptance, the conventional immune response and self-tolerance? These are the big questions in immunoregulation which are the other side of immunology.

Perhaps we should start off with Fas and FasL. In the course of our discussion we considered whether Mike Lenardo's defects in the coding gene segments represent a large part of genetic variation, and decided that they probably don't. If Fas, FasL and the associated signalling molecules are of great importance, as we now think likely, then they should show polymorphic variation in regulatory gene segments. Certainly, a part of the future programme of bringing Fas and FasL into the mainstream of immunoregulation will be through quantitative genetics, with the very powerful technology of quantitative trait loci (QTLs) as the main tool.

Closely linked to that is the issue with which Abul Abbas began the meeting, which is the differential expression of Fas and FasL in different cells of the immune system. Again, we're dealing with the regulation of level of expression, and if it is really so important, then we expect to see that reflected in the quantitative genetics, and we would expect to see associations with immunological diseases.

One might learn more about expression levels from green fluorescent protein (GFP) reporter constructs, which should enable cells with different levels of expression to be isolated by cell sorting. That might help us find out how much these differences in expression level matter. It seems to me that the expression of fluorescent proteins hooked up to the relevant promoters is part of the future.

Lenardo: I think there is a growing sense among geneticists that there will likely be extremely important variations in the level of expression of relevant genes that are part of the susceptibility loci. There are pretty severe limitations on using GFP for studying inducible promoters, because for these studies you want a very rapid read-out that is of the same order of time frame as the induction of the message. In the case of interleukin (IL)-2, you might want it within a few hours. The problem with GFP is that in order to become fluorescent, the newly synthesized protein has to undergo a cyclization reaction that creates the fluorophore that is within the molecule. This usually takes about 18 hours, even for the new mutants that are brighter. This is severely limiting in terms of looking at rapid inducible read-outs. The other technology on the horizon that may well make a big impact here is microchip arrays. In this case you can display 1600 genes on a 1 cm chip. You could make a chip that encompasses these molecules and 1590 others, where we could rapidly look at differences in RNA. You could take a population of lupus patients and do a chip analysis to see whether there's a consistent difference in the expression of a particular gene in the lymphocyte RNA under various conditions among members of a family. In terms of global screening, you can really bring the power of molecular biology into the realm of genetics and autoimmunity.

Mitchison: Isn't the problem with chip arrays that you can't isolate the cells? What we want is to bring expression of these molecules into relationship with immunoregulation. By the time you spread everything out on a chip you can't then use it for anything. You can't sort cells in that way.

Hafler: However, you can take different populations to begin with *ex vivo*, make your cRNA and then use that.

Lenardo: It's just a powerful way of measuring mRNA.

Jenkins: That's a good point, though. I think that we have to begin studying signal transduction in cells right out of the body or *in situ*. Our reliance on cell lines and tissue culture approaches has to be modified so that T cells can be studied as they operate in the lymphoid tissues, not as they operate in 96-well plates.

Allison: One limitation in this system is that Fas display does not indicate a susceptibility to death.

In this meeting we haven't heard anything about the role of Fas/FasL in immunologically privileged sites. What is the current thinking on this?

Abbas: Going back to the original papers, two of the sites that have constitutively high expression of FasL are the testis and the eye, and they happen to be privileged sites. Then transplantation and *in vitro* experiments showed that the testis and the eye will kill Fas^+ targets (Bellgrau et al 1995, Griffith et al 1995). However, Jan Allison and David Vaux have found the experiments with testis grafts difficult to repeat (Allison et al 1997).

Mitchison: Does that apply to the eye also?

Abbas: No. As far as I know, nobody has tried to repeat Griffith and Ferguson's eye experiments (Griffith et al 1995). There is an unrelated sort of story about Fas and FasL which has come out from the laboratories of Lou Matis, Charlie Janeway and others, which says that Fas in the islets is necessary for diabetes induction in the NOD model, but that's probably because it's the target for Th1-mediated cytotoxicity (Chervonsky et al 1997). This has nothing to do with immune privilege.

Allison: In the area of tumour immunity, it has been suggested that many tumours make FasL, so this is going to be a huge problem. Recently we have finished screening our panel of tumours for FasL and found that almost all of them were capable of killing Jarkat cells in the bioassay, but we could blow them away with T cells just fine whether they express FasL or not.

Lenardo: The third report was the report of myocytes expressing FasL creating tolerance to islet cell grafts. I know that Kang and Blau have tried to reproduce this without success.

Abbas: The question you raised is a good one: it is regulated expression and regulated sensitivity, not just structural gene defects or structural gene abnormalities. It is possible that in situations where you don't have a demonstrable structural mutation in Fas or FasL, the pathway may still be defective because of a defect downstream. For example, in the IL-2 knockout mice, the defect may be in Fas-mediated death. The lupus field is full of anecdotal reports of defective IL-2 production or IL-2 receptor expression. Nobody has really put those things together into any cohesive story.

Mitchison: But it's a pretty good test of the importance of that pathway. Either Fas, FasL and the signal-transmitting molecules will show up in QTL analysis of immunological diseases, which is a huge world-wide activity now, or they won't. If they don't, then that is going to place a big question mark over the importance of apoptosis as a regulatory mechanism. It doesn't mean that this system isn't important in other ways, nor does it mean it's not important in setting up the immune system, but as part of the immunoregulation which we're interested in at this meeting, it's got to show up there or we're not going to take it seriously. Would you agree with that Mike?

Lenardo: I'm pretty confident it will show up. It is important to make the distinction that when you look at an outbred population the incidence of homozygosity is low at most loci. If you think about structural problems, they're probably going to be very prevalent because they're going to be dominant interfering in these multimeric complex pathways.

Mitchison: Are you talking about coding sequence variation?

Lenardo: Protein structural alterations require coding sequence changes.

Mitchison: Don't we think that is going to be a minor contribution?

Lenardo: No, it will be a major contribution in the sense of alleles that will be directly associated with diseases. If you are talking about susceptibility loci, where you're envisaging an array of changes that together create a background upon which some environmental change can work, there I think you can talk about gene dosage effects that might have very important effects. There, perhaps variation in message levels could play a much larger role. Thus you're talking about different types of genetic changes that are going to be participating in autoimmune diseases.

Mitchison: Apart from that, there's a certain disappointment: when Fas and FasL first showed up people expected the mutants to show grossly impaired negative selection, but for reasons which are entirely unclear this is not the case. Didn't that come as a surprise to you, Abul?

Abbas: No. I never thought that any one death-delivering molecule would be the key to negative selection, at least in the thymus. It looks like many different signals can trigger apoptosis in immature cells: at least that's what the emerging evidence says. The question that you're really asking is not just about the negative selection in the thymus, but even if you go down to the periphery, if this is such a fundamental way of preventing autoimmunity, why don't you have widespread or multiple autoimmune phenomena? Why are the autoimmune phenomena pretty restricted? Actually, that's a very interesting question. If you believe these reconstitution experiments that are being done putting normal Fas back into various lymphoid cells in *lpr* mice, they would suggest that its dominant role is in T-dependent B cell activation; specifically for eliminating anergic or bystander B cells.

Mitchison: That's where it shows up, but that's not really to do with tolerance induction, it's to do with affinity selection.

Hafler: But if you look at *lpr* (Fas-deficient) and *gld* (FasL-deficient) mice backcrossed on appropriate backgrounds they don't develop experimental autoimmune encephalomyelitis (EAE) (Waldner et al 1997).

Abbas: That was actually *lpr* mice backcrossed onto B10.PL mice, and again the explanation for that has probably got nothing to do with immune regulation. Instead, it is that the major target of Th1 mediated cytotoxicity is Fas, because Th1 cells don't have much perforin and granzyme so they kill via the Fas pathway. Thus if your target cells don't have Fas, then they won't be killed. That is not regulation: it is just effector stage killing.

Mitchison: I wonder if we could clarify the distinction between on the one hand central and peripheral tolerance, and on the other hand between negative selection and other mechanisms such as activation-induced cell death? It is my strong belief that negative selection as it operates in the thymus, operates in precisely the same way in the periphery. Fas/FasL is one of the mechanisms in the periphery, and because it is so important in the thymus, it's likely to be a major mechanism in the periphery. There, it has been difficult to show a role for apoptosis as judged by the genetic analysis which you presented to us using knockouts or transfected additional *bcl* genes. What seems to be clear from that analysis, is that the Fas/FasL apoptosis is likely to be important in another component of immunoregulation which is found only in the periphery: activation-induced cell death and all sorts of other mechanisms.

Simon: What about all the other members of the Fas/TNF receptor family that are being cloned?

Abbas: There is one published result that the CD30 knockout has subtle defects in negative selection in the thymus (Amakawa et al 1996).

Simon: What about DR3?

Abbas: I don't think the newer members of the family have really been examined for their role in the thymus.

Mitchison: Do the apoptosis and developmental groups in Luminy agree that Fas/FasL interactions are unimportant in negative selection?

Simon: We don't believe that Fas plays a role in thymic negative selection, but we think that there may be another member of the family that could play a role. We need double knockouts to be sure. The FADD dominant-negative mouse doesn't have a phenotype in the thymus, but this doesn't mean anything, because there is redundancy.

Mitchison: These are intracellular signalling molecules?

Simon: Yes. The recruitment to the Fas receptor is of course very important. It is not only on the mRNA level, it's everything that happens afterwards.

Abbas: Before we completely dismiss Fas as playing a role in the thymus, there are two published papers, although I have never been convinced by the importance of this pathway in thymic selection. One uses a Fas immunoglobulin fusion protein showing that deletion of quite a few cells in the thymus is Fas dependent (Castro et al 1996). The other is from Jon Sprent, looking at the deletion of semi-mature single positives—these are the heat-stable antigen-high $CD4^+$ cells—where at least under some situations Jon thinks their deletion is also Fas dependent (Kishimoto & Sprent 1997).

Healy: There's actually a third paper from Gerry Crabtree's lab (Spencer et al 1996). Dimerization of a conditional Fas receptor induced apoptosis that was restricted to the double-positive stage of thymocyte development.

Abbas: Most people who have looked at that problem don't really think that it is an important pathway in the thymus.

Lenardo: The problem is that if you look in the thymus for CD95 ligand by PCR, you can't find it, so it's hard to imagine, CD95 (Fas) playing a role.

Mitchison: Let us move onto CTLA-4. It seemed to me that if we want to define better the role of CTLA-4 in immunoregulation, the knockouts aren't going to be much help because they die too soon. The conclusion follows, that the effect of CTLA-4 genetic deletion should be looked at in some kind of transfer system: an adoptive transfer in which CTLA-4^- cells are combined with normal cells in the normal mouse or, better, in a Rag knockout or in bone marrow chimeras.

Stockinger: Would it be possible to cross the CTLA-4 knockout with a T cell receptor transgenic?

Abbas: This is already done, and it cures them of the disease.

Allison: I think you have to be cautious about making that generalization for any single T cell receptor (TCR).

Abbas: With the two CD4 TCR transgenics that we have crossed onto CTLA-4 knockouts (in collaboration with Arlene Sharpe), it lets them live till they are at least 4 or 5 months old. So that is an approach for looking at the behaviour of cells in the absence of CTLA-4, whatever it's going tell you about immune regulation in the more global sense. That is a more difficult question, but at least you can ask what do cells do when they don't have CTLA-4.

Mitchison: I would regard that kind of mouse as a great step forward because it provides cells which will be needed for the combined adoptive transfer experiments.

Abbas: We've put the 3A9 TCR transgene and the DO.11 transgene into CTLA-4 knockouts.

Stockinger: Have you looked at phenomena such as memory and anergy?

Abbas: We have just started looking at *in vitro* responses. The problem is that even in these TCR transgenic CTLA-4 knockouts, the lymph nodes get big with time. The obvious explanation is that they're not on a Rag background. That is a breeding that has started but I'm not looking forward to generating CTLA-4 knockout homozygous, Rag homozygous and TCR transgene-expressing mice.

Allison: We went the other way: we bred the Rag knockout first.

Jenkins: If you are going to do the experiment with the adoptive transfer approach, then the T cell donors will have to be on the same background as the recipients, which will require at least five generations of backcrossing.

Mitchison: Let's move onto the cytokines. The hero of this meeting is IL-10. It has been mentioned repeatedly, and most particularly in David Wraith's and Bernd Arnold's experiments. One has to treat the apparent importance of this cytokine with a note of caution, because its rival, TGF-β, is still not fully accessible to analysis for technical reasons (such as lack of suitable antibodies).

Lechler: It would be better to call IL-10 a potential hero, wouldn't it? There's no data yet either from Bernd or David that it's relevant to anything they've seen in the way of regulation: it is pure correlation.

Mitchison: The question is, how do we take this on from here? FACS staining and so on is still pure correlation. As I understand this message, it's the answer promulgated by Ethan Shevach: you've got to block *in vivo*.

Lechler: And the people who have tried to block, namely Ethan Shevach and Herman Waldmann, have completely failed. It's an open question as to whether cytokine-mediated regulation has anything much to do with what has been discussed at this meeting.

Shevach: However, if one takes for example an SJL mouse, induces EAE, and treats it with anti-IL-10, the disease is enhanced, and the ability of cells to make IFN-γ and transfer the disease is enhanced. So you can see the effects of anti-IL-10 blockade in a normal situation.

Abbas: But that's not the question. Robert's question concerns whether self-regulation is mediated by inhibitory cytokines in a broad sense. Is it correct that there is no direct evidence that suppressive cytokines maintain tolerance to self antigens?

Lechler: I don't know of any.

Mitchison: There's ample evidence that interference with the T cell cytokines affects the autoimmune disease models. For example, anti-IL-4 treatment has dramatic effects in experimental allergic uveitis (EAU) and collagen-induced arthritis (CIA). It is a much more difficult matter to look in the unperturbed immune system at the time of spontaneous induction. I doubt if these cytokines are involved in the initial triggering of autoimmune disease. The general view is that they're more involved in the maintenance and chronicity of autoimmune disease.

Hafler: We have to be cautious. It is clear, for example, that when you administer IL-4 in EAE you can get very different results. There are a lot of surprises with cytokines, which suggests the Th1/Th2 paradigm should not be rigidly adhered to.

Mitchison: Absolutely. But for the moment we're just asking which tools will be needed. Blockade with anti-cytokine antibodies is turning out to be a very powerful (but expensive) tool. We want it available for IL-10 and for TGF-β.

Jenkins: The problem with blockade is that you can't really interpret a negative result.

Mitchison: That's true of science in general.

Jenkins: Disease doesn't happen in germ-free mice, and I don't think anyone has grown an autoreactive T cell out of any of those situations. I think that a failure to regulate a response to commensal microorganisms is just as likely as autoimmunity to be the stimulus for disease.

Mitchison: Once you know that the effects of blockade vary sharply during the course of the disease, you know that knockout isn't going to be very informative, and you know that you've got to use either blockade or conditional knockout.

Hafler: Just getting back to your point on IL-10, when we've looked in individuals infected with chronic Lyme disease, the population that grows out that's reactive to borellia secretes IL-10 and γ-IFN, probably induced by IL-12. Thus, the idea of IL-10 as strictly a Th2-type cytokine is not correct.

Mason: How long will it be before the susceptibility loci for the different autoimmune diseases are actually related to promoter regions for different cytokines? How long is it going to be before we can answer these questions because we know the genetics?

Mitchison: Several of the QTLs which have been identified so far are close to cytokines.

Mason: If you want to ask whether IL-10 is involved in regulating autoimmune disease, if it has a polymorphic promoter then it is a candidate susceptibility gene. How long do we have to wait before we know the answer to that?

Mitchison: You can obtain a negative answer relatively soon provided that the gene in question is polymorphic in the strains chosen for analysis. But if you do see a QTL marker close to the gene in question, you've still got a long way to go before you can identify it. The QTL analysis is now well advanced in EAE (done by at least two different groups), in CIA and in diabetes.

Mason: So where does IL-10 come in this?

Mitchison: So far as I know it hasn't shown up yet. IL-1 is conspicuous, and shows up in all the analyses. It also shows up very clearly in family studies.

Jenkins: The connection between IL-1 and tolerance could again be easily at the level of activating antigen-presenting cells (APCs), specifically dendritic cells (DCs). I could imagine a situation where certain individuals who had hyperactivities in the IL-1 pathway could have a higher general state of APC activation all the time, which could predispose them to autoimmunity.

Mitchison: It's interesting that the family association studies show that kind of activity in almost every group of diseases that you could imagine. The most spectacular recent example is in periodontal disease (Kornman et al 1997).

Abbas: Would most people agree that IL-1 does not have a major action on the majority of mature T cells? If it does play a role in autoimmune disease, it's likely to be indirect via APCs or migration.

Allison: David Raulet's work showed it's required for induction of responsiveness of Th2s to IL-4, but not for the production of IL-4 (Holsti et al 1994).

Abbas: The story on Th2 clones is well established. IL-1 is a growth factor for some Th2 clones. But beyond that, for the vast majority of normal T cell populations, all of us have had trouble demonstrating an effect of IL-1 or IL-1 antagonist.

Shevach: The important point that Marc is bringing up, is rather than using it *in vitro* at all, where the DC functions are maximized because you have purified the cells, you should start injecting IL-1 *in vivo*.

Jenkins: That's exactly what we see. IL-1 has no effect in *in vitro* cultures in activating naïve antigen-specific T cells. It is *in vivo* where it has an effect, which correlates with DC activation.

Mitchison: You told us something important about IL-1, which was that it would activate in your system, but that activation as induced another way was not blocked by anti-IL-1.

Jenkins: We cannot conclude much from our failure to block that effect, because we don't know that the antibody completely neutralized all of the IL-1.

Healy: But if you can show that anti-IL-1 was blocking some other IL-1-mediated event, shouldn't you at least see a partial effect in your system?

Jenkins: Yes, but perhaps that event is more sensitive than the adjuvant event we're looking at.

Healy: Have you seen any effect of IL-1 in any of your controls?

Jenkins: No.

Mitchison: I would certainly argue that the conspicuous IL-1 disease associations have become one of the corner stones of the Zinkernagel view that the activation of APCs is just as important as the decision made by T cells. In that sense, Polly Matzinger has got it right.

Jenkins: And Charlie Janeway had it right before that.

Abbas: Have the QTLs been correlated with actual differences in IL-1 production? That's been a problem with the tumour necrosis factor (TNF) polymorphisms: they were not correlated with TNF production.

Mitchison: The IL-1 polymorphism does correlate with expression (Pociot et al 1994) as does the TNF one (Messer et al 1991). It is very clear. I would emphasize, by the way, that showing differences is only the beginning of the genetics. What you have to account for is selection for heterozygotes. So just differences in level can't be the key: it's difference in expression at different places.

Let's move on then to the phosphates, CD45 and Shp-1. If I understood what you were really getting at in your paper, Richard Cornall, it was the possibility that anergy, in this case in B cells, reflects phosphatase activity. A cell may be anergic because of a high level of phosphatase, which is blocking signalling.

Cornall: We don't know what generates the proximal block of signalling in anergic cells. It is our increasing understanding that the state of anergy is actively controlled by through pathways. This is the work of Jim Healy in Chris Goodnow's lab. But we haven't discovered the cause of the proximal block.

Healy: We do know that in the tolerant cells some of the selective signalling which I described is dependent upon CD45 either for its induction or its maintenance. Although genetically Shp-1 is clearly a modulator of B cell receptor signalling, we have been unable to obtain biochemical evidence to suggest that it's involved in the modulation of signalling in the tolerant B cells.

Allison: Do we accept that CD22 and maybe CTLA-4 are involved in setting the thresholds for signals that are required for initial activation and tuning? They may very well fall out of the sort of genetic screen that you are talking about.

Healy: Yes, it would be interesting if they did.

Mitchison: The cross-linking studies show that if you bring CD45 into the TCR complex it can silence it.

Shevach: That is what is published.

Allison: But I think that one needs to consider CD45 as a positive phosphatase in that sense: it keeps everything primed. The negative phosphatases are more interesting.

Mitchison: I would like to ask the same question I asked about the Fas: is it not possible that hooking up GFP to the promoters of these phosphatases might be revealing?

Abbas: Are you suggesting attaching GFP to the promoter region of the phosphatase and then looking for transcriptional activation of the phosphatase as a marker for anergic cells? The reason that may not work is that I don't think phosphatase activity is regulated by transcription.

Allison: An experiment we have on the books that relates to that, is to try to make a GFP fusion with CTLA-4 or CD28, and then try to correlate traffic towards the surface with levels of calcium in the cell.

Mitchison: I agree that the rapid signalling events must reflect activation, not transcriptional control. Nevertheless, regulation of transcription is likely to be the main way of controlling long-term cell activity, as in anergy.

Healy: Transgenic mice with an AP-1 promoter linked to a luciferase reporter gene (Roncon & Flavell 1994) induce AP-1 transcriptional activity in T lymphocytes only after anti-CD3 stimulation for two days. This stands in contrast to Jun kinase activation which occurs in minutes and c-*jun* induction which occurs within an hour. Reporter gene transgenics are not good models for signal analysis.

Jenkins: Despite this, I think that's a place we need to start going. We need to see whether AP-1 driving GFP fails to fire in cells that persist after a tolerization protocol *in vivo*.

Lenardo: Actually, better experiments would be where you can knock-in a reporter into the gene locus and monitor how that gene is firing in different states of the cell.

Healy: That may be more powerful.

Allison: One of the problems here is that many of the molecules that are involved are already there: they are not regulated by transcription, but rather by kinase and phosphatase cascades.

Healy: One still wants real time biochemical analysis since the kinetics of activation and inactivation are clearly important.

Allison: You can use these techniques, but I think you have to use them *in vitro*.

Mitchison: All a transcription assay tells you about is transcription, which may or may not be important. You don't know enough to say that it's not important.

Allison: With these kinases and phosphatases, I think we probably do. The proteins are there. That is always the baseline for the measurement of pre-activated or activated cells: you immunoprecipitate and see whether there is the same amount of protein there before you infer anything.

Allen: The point is that in many of these systems where the phosphatases are important, the system appears to be primed to be activated and the phosphatases keep it in an inactive state. In this way the system responds much quicker, and any transcriptional events are going to occur much later.

Mitchison: If we're trying to look for mechanisms of anergy, we're not in such a hurry. There we're looking for something that's going to last for a few weeks at least.

Cornall: Yes. Whatever establishes anergy must be able to establish a steady state for several weeks, provided the cells remain in contact with self antigen.

Abbas: One other approach that many people have been trying is differential display PCR for various populations of anergic and functional cells. It has not led to very much yet, but there is always hope.

Mitchison: Part of the future surely lies in scanning the genome for promoter recognition elements occurring in combination; PREG-scanning as I call it.

A remark about animal experiments. It's my belief that sooner or later we're going to have to go back to experiments on fairly large numbers of mice, in spite of the expense, bureaucracy and problems of interpretation (in contrast to 96-well immunology) which it entails.

Abbas: In fact, genomics companies realize that after identifying disease-associated DNA sequences, we will need to examine the functions of these sequences *in vivo*. This will cost a fortune.

Mitchison: The last question I wanted to raise seems to have arisen several times during this meeting, and we should air it again now. It concerns activation of protective T cells. Can't we do better in raising protective T cells then simply feeding peptides or proteins or having animals sniff them? For decades the use of adjuvants has been regarded as anathema in the context of inducing tolerance. If you activate the immune system with adjuvants, this stops tolerance from being induced. Perhaps we should backtrack on that now, and be thinking about appropriate adjuvants for use, for example, in activating the protective cells which we've heard about today from Ethan Shevach. By the term 'adjuvants' I include drugs which have selective action in inducing Th1 or Th2 cells. The route of immunization is also important. We've got to find out how to arouse these protective cells.

Wraith: I agree wholeheartedly with that. If the induction of the tolerant state that we're seeing is an active process, then it could well be that the use of adjuvant in some form will be incredibly helpful. I can tell you without any details of one such experiment where one can tune down the amount of antigen that you are feeding for oral tolerance by linking to the β-subunit cholera toxin.

Hafler: Absolutely, the cholera toxin B chain is a very potent adjuvant, as are IL-4 and IL-10. Antigen-specific immunotherapy is not that complicated: there are only a certain number of ways of administering antigen, and then there may be certain adjuvants to enhance it.

Shevach: I am not enthusiastic about the use of adjuvants to generate antigen-specific tolerance. The problem with adjuvants is that they are very likely also to induce inflammation. The presence of a small amount of IL-1 or lipopolysaccharide (LPS) could easily lead to induction of co-stimulatory molecules or the production of IL-12 which would promote a Th1 response. However, Fred Finkelman and colleagues have devised a system where antigens can be targeted to dendritic cells in the absence of an inflammatory response with the induction of specific tolerance (Finkelman et al 1996).

Mitchison: I wonder if you're right about that. After all, every other T cell needs a bit of inflammation to get itself going. Maybe at the right time, the right place and by the right route, a little bit of inflammation is exactly what your protective cells need.

Shevach: It is amazing that the IL-12 knockout does not develop a Th2 response when immunized with protein antigens in complete Freund's adjuvant.

Jenkins: The key point here is that 'other cell' part. Are these suppressor cells really a different lineage of cells, or are they different fates that a single naïve T cell can acquire? As I sat here and listened to the confusion in all these different systems, my sense is that we don't understand memory cells. We have very simple ideas that a naïve T cell becomes a Th1 or Th2, or anergic or dead. I think there is much more heterogeneity within memory cell populations than this. We need to understand the possible functional properties that memory cells can acquire, because I'm not convinced that the 'suppressor' cell is a different lineage of cells. It may be a specialized memory T cell.

Hafler: There's nothing to suggest that it is.

Jenkins: That is where adjuvants could play a role, in determining how far down the memory cell pathway the cell is going to go. If IL-4 or IL-12 is there, this is definitely going to have an effect on how far they can go. If these cytokines are absent, then maybe 'suppressive' memory cells are generated.

Wraith: We generate the same cell from the same transgenic mouse.

Jenkins: But nobody has really shown that this cell can suppress anything, as far as I can see. We did a direct test of that and didn't find any evidence for suppression. No one has shown that a T cell with a defined specificity can suppress another T cell in a way that's unambiguous, for example in a TCR transgenic system. If we look at the other mechanisms of tolerance, e.g. deletion and anergy, TCR transgenic mice or some other method of tracking the antigen-specific cells of interest has been used to validate the mechanism. That's where the suppressor field has had problems in the past.

Mitchison: You know that by now that the transgenic will make every other kind of T cell: how is it possible that it could not make a suppressor cell if such a cell exists?

Jenkins: I think we have no clue about that until someone identifies a suppressor cell and makes the transgenic mouse.

Mitchison: To close, let us remind ourselves that we need to deliver something of therapeutic value. Surely the point of studying tolerance is to find out how to prolong allograft survival and suppress autoimmune disease. The success of this meeting, in the long run, will be judged by how much it contributes to achieving that aim.

References

Allison J, Georgiou HM, Strasser A, Vaux DL 1997 Transgenic expression of CD95 ligand on beta islet cells induces a granulocytic infiltration but does not confer immune privilege on islet allografts. Proc Natl Acad Sci USA 94:3943–3947

Amakawa R, Hakem A, Kundig TM et al 1996 Impaired negative selection of T cells in Hodgkin's disease antigen CD30-deficient mice. Cell 84:551–562

Bellgrau D, Gold D, Selawry H, Moore J, Franzusoff A, Duke RC 1995 A role for CD95 ligand in preventing graft rejection. Nature 377:630–632

Castro JE, Listman JA, Jacobson BA et al 1996 Fas modulation of apoptosis during negative selection of thymocytes. Immunity 5:617–627

Chervonsky AV, Wang Y, Wong FS et al 1997 The role of Fas in autoimmune diabetes. Cell 89:17–24

Finkelman FD, Lees A, Birnbaum R, Gause WC, Morris SC 1996 Dendritic cells can present antigen *in vivo* in a tolerogenic or immunogenic fashion. J Immunol 157:1406–1414

Griffiths TS, Brunner T, Fletcher SM, Green DR, Ferguson TA 1995 Fas ligand-induced apoptosis as a mechanism of immune privilege. Science 270:1189–1192

Holsti MA, McArthur J, Allison JP, Raulet DH 1994 Role of IL-6, IL-1 and CD28 signaling in responses of mouse CD4+ T cells to immobilized anti-TCR monoclonal antibody. J Immunol 152:1618–1628

Kishimoto H, Sprent J 1997 Negative selection in the thymus includes semimature T cells. J Exp Med 185:263–271

Kornman KS, Crane A, Wang HY et al 1997 The interleukin-1 genotype as a severity factor in adult periodontal disease. J Clin Periodontol 24:72–77

Messer G, Spengler U, Jung MC et al 1991 Polymorphic structure of the tumor necrosis factor (TNF) locus: an NcoI polymorphism in the first intron of the human TNF-β gene correlates with a variant amino acid in position 26 and a reduced level of TNF-β production. J Exp Med 173:209–219

Pociot F, Ronningen KS, Bergholdt R et al 1994 Genetic susceptibility markers in Danish patients with type 1 (insulin-dependent) diabetes—evidence for polygenicity in man. Danish study group of diabetes in childhood. Autoimmunity 19:169–178

Rincon M, Flavell RA 1994 AP-1 transcriptional activity requires both T cell receptor mediated and co-stimulatory signals in primary T lymphocytes. EMBO J 13:4370–4381

Spencer DM, Belshaw PJ, Chen L et al 1996 Functional analysis of Fas signaling *in vivo* using synthetic inducers of dimerization. Curr Biol 6:839–847

Waldner H, Sobel RA, Howard E, Kuchroo VK 1997 Fas and FasL deficient mice are resistant to induction of autoimmune encephalomyelitis. J Immunol 159:3100–3103

Index of contributors

Non-participating co-authors are indicated by asterisks. Entries in bold indicate papers; other entries refer to discussion contributions.

Subject index

D

J

K

L

M

V

X